SCALING SUPPLY CHAINS WITH MARIA

Scaling Supply Chains With *Maria*

Financial and Operational Frameworks and Analytics for Massive, Profitable Growth

Marcia D. Williams

Scaling Supply Chains with Maria © 2023 Marcia D. Williams

All rights reserved. No part of this publication may be reproduced, distributed or transmitted in any form or by any means, including photocopying, recording, or other electronic or mechanical methods, without the prior written permission of the publisher, except in the case of brief quotations embodied in critical reviews and certain other noncommercial uses permitted by copyright law.

Although the author and publisher have made every effort to ensure that the information in this book was correct at press time, the author and publisher do not assume and hereby disclaim any liability to any party for any loss, damage, or disruption caused by errors or omissions, whether such errors or omissions result from negligence, accident, or any other cause.

Adherence to all applicable laws and regulations, including international, federal, state and local governing professional licensing, business practices, advertising, and all other aspects of doing business in the US, Canada or any other jurisdiction is the sole responsibility of the reader and consumer.

Neither the author nor the publisher assumes any responsibility or liability whatsoever on behalf of the consumer or reader of this material. Any perceived slight of any individual or organization is purely unintentional.

The resources in this book are provided for informational purposes only and should not be used to replace the specialized training and professional judgment of a health care or mental health care professional.

Neither the author nor the publisher can be held responsible for the use of the information provided within this book. Please always consult a trained professional before making any decision regarding treatment of yourself or others.

For more information, email marcia.williams@usmsupplychain.com

ISBN: 978-1-7343815-3-5

This book is dedicated to the following amazing persons:

- You, the reader, who has the grit and mindset to make a difference through supply chain and procurement.

- My grandma, Otta, whose love and actions have taught me the most powerful lessons. She is in heaven, but also present in my heart.

- My mom, who has always believed in me. Her encouragement, perspectives, and optimism have been inspiring. I couldn't ask for a better mom.

- My dad, who could not have supported me better through the crazy paths I have taken. His hard work, commitment, and perseverance are exemplary.

- My husband and children, who have created a wonderful world for me. They help me to be a better person each day.

- My sister, who is always willing to help. Her courage and strength are incredible. She is the proud mom of my sweet niece and nephew.

- My aunt and her family, who have cherished and cared for me.

- Americo and Antoniette D'Addonna, whom I consider to be like parents. I have learned so many valuable lessons from them. Their help has been outstanding. I love them dearly.

- Rose Di Giglio, an awesome mom whose love, endurance, and enthusiasm are an example to follow.

RESOURCES PAGE

To live the full experience of Scaling Supply Chains with Maria, make sure to visit the resource page at **usmsupplychain.com/resources/** or scan the QR code above.

You will find:

Excel Tools & Templates

Summary Videos

Focus Areas

Key Notes

Visuals

Table of Contents

Introduction 13
Chapter 1: The Challenges 17
   The Top 10 Supply Chains Challenges for Alex's Snacks 17
      Greater Sales; Lower Margins 18
      Higher Packaging Costs 18
      Higher Shipping Costs 19
      Warehousing Issues 19
      Manufacturing Challenges 20
      Oil and Other Material Shortages 20
      Excess Inventory 21
      Lost Sales and Late Deliveries 21
      Lower Fulfillment Rates 22
      Delayed New Product Launches 22
   The S.C.O.P.E. (Superhero Challenge of Processes Expedited) of Work . . . 23
Chapter 2: The Powerful Impact of Supply Chain on the Financials . 27
   Key Takeaways 27
   Financial statements: Profit and Loss Statement (P&L), Balance Sheet, and Cash Flows 29
      Cash Basis vs Accrual Accounting 29
      Cash Flow Statement 32
      P&L 33
         Revenue 34
         Cost of Goods Sold 35
         Gross Profits 37
         Operating Expenses 37
         Operating Profit or Operating Income 37
         EBIT 38
         Other Income 39
         EBITDA 39
         Interest and Taxes 40
         Net profits—Bottom line 41
      Balance Sheet 41
         Assets 42
         Liabilities 43
         Equity 45
   Five FAQ—Frequently Avoided Questions—on Supply Chain and Procurement Impact on the Financials 46
      Purchasing In Larger Quantities 47
      Not Having the Material On Time 50

    Shipping in Larger Quantities 52
    Payment Terms Extension to Suppliers 52
    Liabilities and Accruals. 53

Chapter 3: Metrics 55

Key Takeaways 55
Finance-Supply Chain Alignment on Key Metrics 55
    Liquidity 55
        Current Ratio 56
        Quick Ratio 56
        Operating Cash Flow Ratio 57
    Debt or Solvency 57
        Debt to Equity Ratio 58
        Equity Ratio 58
        Debt Ratio. 58
    Profitability 59
        Gross Profit %. 59
        Operating Profit %. 59
        Net Profit % 59
    Operating Performance, Efficiency, or Activity Ratios 60
        Operating Efficiency Ratio 62
        Accounts Receivable Turnover Ratio 62
        Average Collection Period 62
        Accounts Payable Turnover Ratio 62
        Average Number of Days Payables Outstanding 63
        Inventory Turnover Ratio 63
        Days Sales in Inventory 64
        Asset Turnover Ratio 64
        Fixed Asset Turnover Ratio 64
        Total Assets to Sales 65
        Fixed Assets to Total Assets. 65
        Working Capital Ratio 65
    Investment Valuation or Market Ratios 66
        Price to Earnings ratio 67
        Price/earnings to growth ratio (PEG ratio) 67
        Price to sales ratio 67
        Price to book value ratio 68
Metrics in CPG for benchmarking 68
    Common Metrics and Levels in CPG. 70
        Profit Margins 71
Strategic Goal Tree and SCOR model 73
    SCOR DS Model 76
    The WHY of SCOR 77
    The WHAT and HOW of SCOR 78

- Structure of the SCOR framework 79
  - Performance 79
  - Processes 79
  - Practices 79
  - People 79
- Our Beloved Performance Metrics 79
  - Resilience 80
  - Economic 80
  - Sustainability 80
- Processes in SCOR 85
- Practices in SCOR 86
- SCOR DS Model Limitations 87

Chapter 4: Process Mapping, Matrices, Frameworks, and Toolkit from Lean Six Sigma 89

- **Key Takeaways** 89
- **The Why of Lean Six Sigma and Fundamentals** 89
  - Six Sigma 90
  - Lean Method 90
- **Lean Six Sigma Toolkit** 93
- **The 5-step DMAIC Approach (Define, Measure, Analyze, Improve, Control)** . 94
  - Define 95
    - Six Sigma Project Selection Process 96
    - Six Sigma Project Description 98
    - Funneling Effect 100
  - Measure 103
    - Process Mapping 103
    - 50,000 ft. and 5,000 ft. Process Variables Map 109
    - Instructions to Complete the 50K ft. and 5K ft. Process Maps in 4 Steps 109
    - C&E 111
    - Instructions to Complete the Cause & Effects Matrix (C&E) in 5 Steps 114
    - FMEA 115
    - Instructions to Complete the Failure Mode & Effects Analysis Matrix (FMEA) in 9 Steps 119
    - FTA – Fault Tree Analysis 121
    - 5 Whys—the "Toddler Approach" 125
    - MSA 126
  - Analyze 128
    - Ishikawa Diagram—Fishbone Diagram—Cause and Effect 128
    - Scatter Graphs 129
    - Pareto Charts 130
    - Non-Value-Added Analysis 131
    - t-Test 131
    - How to Perform a T-test Step-by-Step 133

  ANOVA 134
  Chi-Squared Test 135
  Design of Experiments (DOE) 136
  Instructions for DOE in 5 steps 137
  Taguchi 137
  Regression Analysis 138
 Improve 140
  Affinity Diagram—"Post-brainstorming With Order" 140
  How to Create an Affinity Diagram in 4 Steps 143
  K-J Analysis 144
  5S—"Marie Kondo" 146
  Poka Yoke—Mistake Proofing 148
  Poka Yoke in 5 Steps 148
  Kaizen 150
  Kanban 151
  Kanban Boards and Cards 152
  Kanban Cards—Kan = Card; Ban = Signal 153
  Pugh Matrix 154
  Heijunka = Leveling 155
  SMED—Single Minute Exchange of Dies 157
  Total Productive Maintenance (TPM) 158
 Control 159
  PDCA (plan, do, check, act) 159
  Standardized Work 160
  FMEA 161
The 5-step DMADV Approach. **161**

Chapter 5: S&OP (Sales & Operations Planning)/IBP (Integrated Business Planning); ERP; MRP 165

Key Takeaways **165**
S&OP and IBP **166**
Financial Integration with S&OP to Move Towards IBP **171**
IBP Dimensions, including Finance, Demand, New Products,
Supply, and Executive **173**
Demand Planning **173**
 Statistical Forecasting 173
 Qualitative approaches 174
  Grassroots approach 174
  Market research 175
  Executive judgment 175
  Historical analogies 175
  Delphi Method 175

- Quantitative approaches 176
  - Causal forecasting 176
  - Time series forecasting 179
- Selecting the Right Forecasting Technique 184
  - Purpose of the forecast and desired accuracy 184
  - Cost. 189
  - Availability of historical data 189
  - Availability of forecasting software 192
  - Time needed to gather and analyze data and prepare a forecast 197
  - Forecast horizon 198
  - Life-cycle stage of forecast for a particular product 198
- Product Portfolio Management; SKU Rationalization; Product Portfolio Optimization 199
  - Key Areas and Challenges 199
- Trade Promotion Management 204
  - TPM in 5 Steps 206
- Demand Sensing 206
- Demand Shaping 207

Supply in Depth with a Holistic Approach, Considering the Impact on the Financial Statements 210

Inventory Management and Optimization 210
- Supply chain and finance together, forever and ever 210
- Inventory 211
- Make to Stock (MTS) and Make to Order (MTO) 212
- Impact on Financials 214
  - Cost Accounting 214
- Standard Costing and Activity-based (ABC) Costing 216
- Inventory Transactions (Normal and Standard Accounting) 217
  - Raw Material Purchases 219
  - Interest 221
  - Inventory write-off 222
  - Raw Materials Receipts with No Invoice 224
  - Payment to Raw Material Suppliers 226
  - WIP Creation 226
  - Finished Goods—Assemblies 231
  - Finished Goods Sales 232
  - Inventory Transfers 234
  - Finished Goods Returns 234
- ABC Analysis 235
- ABC Step-by-Step Instructions 238
- What to buy, what to produce, how much, and when 240
  - Economic Order Quantity 244
  - Safety Stock 246

The Baker's Dozen Metrics For Inventory Management 249
   Inventory Days of Supply—Assets. 249
   Cash-to-Cash Cycle Time—Assets 250
   Return on Working Capital—Assets 251
   Inventory Carrying Cost—Cost 252
   COGS—Cost 252
   Total Obsolescence for Raw Material, WIP, and Finished Goods Inventory—Cost. .253
   Perfect Customer Order Fulfillment—Reliability 253
   Forecast Accuracy—Reliability 254
   Inventory Accuracy—Reliability 254
   Customer Order Fulfillment Cycle Time—Responsiveness 255
   Source Cycle Time—Responsiveness. 256
   Fulfill Cycle Time—Responsiveness 257
   % of Renewable Materials Used—Sustainability. 257

Procurement and Procure to Pay, extending into Logistics 258
   Total Cost of Ownership 261
   The Baker's Dozen Effective Ways to Achieve Cost Savings—
   Examples with Logistics and Fulfillment Included!. 263
      Ask suppliers for cost reduction. 263
      Better Planning 267
      Supply Chain Network Optimization. 271
      Right Sizes in Buying, Manufacturing, Shipping 276
      Process Optimization 277
      In-house vs. Outsourcing 278
      MBE (Minority Business Enterprise) Suppliers. 280
      Supplier Segmentation 281
      Quick Requests for Quote. 284
      RFXs (RFI, RFP, RFQ), including Online Auctions 284
      Contract Management 288
      Changes to the Packaging and/or Product 291
      Opportunities in Procure-to-Pay. 292

Introduction

Is your company experiencing exponential growth, either organically or by acquisition? With skyrocketing sales, your supply chain may not keep up with the pace, hindering your on-fire growth and slashing your profits. The business did an amazing job in scaling up sales, and now your supply chain and procurement team feels a sharp and acute growing pain with:
- Lower profit margins
- Increased costs
- Lost sales
- Late customer deliveries
- Delayed new product launches

It's time for your supply chain to step up to meet the frenetic and unpredictable rhythm of sales—like a horse galloping away in the wilderness. This book equips supply chain and procurement practitioners with the frameworks, processes, and analytical tools to take the reins and deliver stellar financial results. This rodeo is going to be fun too!

Scaling Supply Chains with Maria takes place in a fictional potato and tortilla chip manufacturing company that is going through an explosive growth of its retail and e-commerce sales, like an unstoppable volcano eruption—causing all sorts of pain to an already worn-out supply chain and procurement team because of Covid-19, the Ukraine-Russia war, and other disruptions.

Maria and the team members—the characters in this book—focus on tying the financial and operational aspects of the business. Chapters one, two, and three relate to understanding the impact of supply chain and procurement on the financial statements and mapping out a strategic goal tree with metrics that align business strategy, finance, and supply chain by applying the SCOR framework.

In chapter four, Maria presents a lean six sigma toolkit with analytical tools that the team executes to address the supply chain and procurement challenges. The book covers the DMAIC framework for continuous improvement as well as DMADV for new products.

The final chapter is about building an authentic Integrated Business Planning (IBP)—driven by financial targets and starting with the financial forecasts of the organization. This chapter also covers supply chain optimization—including inventory—considering the end-to-end supply chain. Maria introduces a toolset for demand forecasting, analytics, and simulation or modeling. Towards the end of the chapter, the focus is on procurement and how to achieve effective cost reduction.

These financial and operational frameworks work and do a lot of the work for you. I'm telling you this, based on almost two decades of experience in finance-supply chain adventures; of course, I started when I was five. Now seriously, I have had the privilege to work with high-profile companies—many Fortune 500—including Hershey, Lindt Chocolates, Coty, Cummins, and Alcoa. Below is what *they say* about my performance that you can validate on LinkedIn or you can Google me:

"Marcia is a highly skilled professional in Supply Chain; her support has been key to overcome critical aspects of our Procurement Operation. We work together developing processes and tools; with her support we were able to automate complex processes transforming them through easy-to-operate tools that have given us accurate and reliable results. Her commitment to our Company was remarkable; Marcia has a great willingness to share her knowledge; it was a real pleasure to work and to learn from Marcia" **Jose Antonio Laparte, Sr. Procurement Manager, Salty Division at the Hershey Company**

"Marcia brought a high level of expertise and diligence to our project at Lindt, helping to define and validate requirements for our supply-chain planning project. Her experience proved invaluable to quickly deriving, researching, and making recommendations as to how we would proceed in our software selection process, while using existing materials and creating her own. I would highly recommend Marcia for other supply chain projects, especially for CPG or other manufacturing clients." **John Vaughan, Business Systems Analyst Principal at Allianz Investment (former IT Data Analyst at Lindt Chocolates)**

"Marcia is results-driven, strategic, and focused on cost savings. She has strong, practical experience with Six-Sigma tools and processes to receive valuable results to the bottom-line. She is a pleasure to work with and willing to do whatever it takes." **Mark Farney, Talent Acquisition Leader at Cummins Inc.**

The frameworks, processes, and tools that I've included in this book are critical for result-oriented and data-driven supply chain and procurement teams. They will help you successfully navigate the rough waters of fast growth—either organically or by the acquisition of other brands and companies—to achieve massive financial results.

I have written this book in a storytelling format with characters, breaking down complex concepts and analytics in finance and supply chain to make it simple and fun to understand. At the same time, I have made it highly actionable, coming from my accounting background and supply chain experience in the trenches doing the actual work like you do. So, strap in and buckle up: Read this book, take action, and you'll be able to deliver high impact results!

About the Series

The Series Supply Chain with Maria uses stories with characters in fictional companies to turn complex concepts into simple and actionable to deliver insane financial results through supply chain and procurement.

"One who sweats more in training bleeds less in war"

Spartan Credo.

CHAPTER 1:

The Challenges

"Good morning; is this Maria?"

"Hello, it's me," Maria answers her phone following Adele at her best in her song "Hello," but without the singing.

"I'm Ernesto Gonzalez, VP of Supply Chain and Procurement at Alex's Snacks. I've got your contact information from The Walls. They were very happy with your performance at their company, The Long Island New York Chocolates."

Maria interjects, "Of course, I remember Tom and David and all the delicious candy I had during our supply chain transformation journey."

Ernesto chuckles and adds, "I think you can help us with some challenges we are experiencing here in supply chain and procurement, as we are also in the consumer-packaged goods (CPG) industry."

As top executives do, Ernesto takes immediate action and sets up a meeting through Microsoft Teams the next day with Maria.

The Top 10 Supply Chains Challenges for Alex's Snacks

During the conversation, Ernesto starts with an overview of his business, located in South Central Pennsylvania. Many consider this area the snack food capital of the world. It's a couple of hours away from Philly.

The humble beginnings of the business are with a great recipe treasured for over a decade by now. Products include potato, tortilla, and kettle cooked chips. The annual sales are around $200M. Management expects nothing but more growth, at the pace of a rocket launched to the moon. Alex's Snacks is hot and on fire to become a prominent leader in the snack industry.

The conversation takes a turn to the challenges in supply chain and procurement that are the company's major enemies in accomplishing more and more profitable growth. This turn also stops Maria's unavoidable mouth-watering with the flavored chips' images.

Like the first brushes of a painter, Ernesto reveals the laundry list of supply chain and procurement pain points that keep him up at night and make him beat the sunrise with no alarm or coffee. They are as follows:

Greater Sales; Lower Margins
Ernesto feels that "supply chain is like the Pac-Man in older video games eating our profits." Sales don't equal profits. Gross pay is not the same as net pay. What matters is what the company can reinvest or distribute or what the employee can take home.

Alex's Snacks is having a feast with greater revenues. At the same time, the margins are becoming slimmer. It is like being on a diet during Thanksgiving and the Holidays when there is plenty of food (or sales), but the portion the diner can take represents a lower percentage.

Higher Packaging Costs
Covid-19 spreads an assortment of supply chain issues, including raw material shortages for film packaging manufacturers. These shortages have an impact on cost and lead times. Cost goes up about 40% for Alex's Snacks. Lead times are longer, from eight weeks to 16 or up to 20 weeks in some instances.

Corrugated packaging follows a similar pattern. Cost goes up 35%. Lead times go from one or two weeks to three to four weeks. Boxes and shippers can become precious!

Both film and corrugated packaging increased cost slashes Ernesto's company profitability. Ouch!

Higher Shipping Costs

"I know how it feels to get a driver for a Friday delivery," Maria shares with Ernesto. "It is as easy and pleasurable as bathing a cat," as the writer Anne Lamott would say.

The American Trucking Association calculates the truck driver shortage as the difference between the number of drivers in the market and the optimal number of drivers based on freight demand. The gap remains at near-record high[1]. This is a key cost driver for the longer haul.

The pandemic has caused havoc in the rates and timelines of ocean and air freight. There were times when if a company shipped ocean, they didn't know when the shipment would arrive or if it would arrive someday…This situation is changing back to normal, but it has made shipping rates soar.

Expedited shipping is another factor hurting profitability. Ernesto shakes his head with the countless times that his team expedites shipping to meet the schedule. More money thrown at the "profit Pac-Man" supply chain.

Warehousing Issues

Alex's Snacks doesn't own a warehouse. The company relies on a 3PL for warehousing of their finished goods—the insanely good chips ready and packaged to get to consumers' hands. For raw materials—ingredients and packaging—the business holds inventory at their co-manufacturers' and co-packers' locations and some at the suppliers' sites.

There are space constraints everywhere. It feels like being the prey of a boa constrictor with little room to breathe because of the inventory pile. More inventory. No place to put it.

Manufacturing Challenges

Production takes place at the co-manufacturers' plants. Ernesto shares with Maria that there have been thoughts of bringing manufacturing in-house to have a tighter control of operations.

Alex's Snacks follows two current business models with its co-manufacturers. The main difference lies in the raw material acquisition. The business models are as follows:
1. *Contract Manufacturing*—This is all-in. The co-manufacturer produces the chips and acquires the ingredients and packaging. The co-manufacturer owns the inventory.
2. *Tolling Manufacturing*—With this model, the co-manufacturer also produces the chips and Alex's Snacks buys the raw materials. Alex's Snacks owns the inventory.

"I bet that production at the co-manufacturers is like opening a Pandora's box," says Maria. She doesn't like "the can of worms" idiom. It sounds kind of nasty to her. Ernesto mentions production lines down, changes to production schedule, rejection of incoming material, quality issues, communication with the team, among other issues or worms, because this is nasty.

Oil and Other Material Shortages

Shortages hit the business during Covid, causing severe supply chain disruptions and raw material price hikes. Then the Ukraine-Russia war smashed the company's COGS (cost of goods sold) with the price spikes of cooking oil; mainly the sunflower and cottonseed variety needed for the chips' production. Alex's Snacks faces outrageous cooking oil prices and not enough oil quantity or supply to meet the company's demand. It seems like Ernesto's "profit Pac-Man" likes them fried.

Excess Inventory

Too much inventory is a big challenge at Alex's Snacks in the Covid-19 aftermath. From supply chain disruptions to inventory floods, giants like Nike and Walmart are also sharing this common profound pain.

For the well-known sportswear company, the in-transit inventory increase makes inventory levels jump to 65% over Q1 of the previous year[2], when factories in China and Indonesia shut down.

Nike had placed early holiday orders, anticipating shipping delays. It turns out that in-transit times have improved and now the "just buy it" company sees inventory piles up.

Walmart has a similar situation with inventory levels pushed up 35% against Q1 of the prior year. News sites[3] describe the excess inventory situation at the retailer as "chaotic, overcrowded backrooms, and outdoor storage units stuffed with unsold goods." Others go further with "Walmart's excess-inventory issues are spilling out into its stores."

Within this context, Maria shows no surprise when Ernesto indicates that they are sitting on piles of inventory, ranging from potatoes to film. These inventory issues feel like a big punch—even for Stallone and Schwarzenegger—and non-stop punches at profits!

Lost Sales and Late Deliveries

Lost sales are unmet demand. Alex's Snacks receives more customers' orders than what the company can fulfill. There is a gap between demand and sales.

This gap occurs because of a myriad of reasons; constrained production capacity, material shortages, longer lead times, shipping delays, changes in customer demand and other factors.

Lost sales play havoc on the company's fast growth as invisible frost damages the sprouts in trees. Finance doesn't see lost sales as a top line in the P&L (profit and loss) statement, but they do exist at Alex's Snacks.

The company's lost sales are leaving millions of revenues—and profits—on the table. They can lead to lost customers, too. Per Ernesto's explanations, Maria understands that this is real and a valid concern. A subtle and quiet way for Ernesto's supply chain Pac-Man to eat more of the company's profits.

Late deliveries to big accounts are also a big headache for Ernesto. The C-Suite keeps asking him for explanations. Ernesto feels the frustrations from the customers and from management at Alex's Snacks.

Lower Fulfillment Rates

Alex's Snacks has challenges with the fulfillment of orders. Covid has a negative impact, but it is not the sole cause. A key driver of lower fulfillment rates is the business's fast growth.

Supply chain can't keep up with customer demand and the added complexity. They are dancing to different songs. Demand is a faster dancer than supply chain. Supply chain can't follow the demand's pace. It is way too fast.

When supply chain attempts to be in sync with demand, there are missteps that result in lower fulfillment rates that could lead to lost sales and lost customers. At Alex's Snacks, it looks like these two, demand and supply chain, need to learn how to dance together to shine on the shop floor.

Delayed New Product Launches

New product launches are fundamental for Alex's Snacks, as they are for every CPG company. CPG businesses introduce new products to meet the ever-increasing customer requests for variety.

According to Nielsen IQ's research[4], the CPG industry launches an average of 30,000 new products each year. The same analysis indicates that one-third of Americans are actively looking for new products to try.

This trend extends post-Covid as consumers continue to ask to spice things up.

The Challenges

Persistent as a toddler asking to play Roblox or Minecraft, inventory issues jeopardize new product launches. Material shortages, longer lead times, and regulatory and compliance prevent Ernesto's business from introducing amazing new products per the planned schedule.

The other side of the coin—excess inventory of the parts set for discontinuation—plays a critical villain role. Maria can hear the silent screaming of Ernesto, "When are these dark clouds going to pass? Will we be able to find a pot of ridiculously good potato chip launches (and profits) at the end of the rainbow?"

The S.C.O.P.E. (Superhero Challenge of Processes Expedited) of Work

After the conversation with Ernesto, Maria understands that Alex's Snacks is looking for a S.C.O.P.E. of Work. S.C.O.P.E. meaning Superhero Challenge Of Processes Expedited.

This Superhero challenge is for his entire supply chain and procurement team with Maria's leadership and support. His team needs processes, frameworks, and tools to accompany the fast growth in sales, and scale profits along with the supply chain. It is a whole new level of playing field that requires his organization to become more professional to perform and succeed.

"There is no time to waste," Ernesto emphasizes. There is a call for speed—the "expedited" component of the challenge.

Ernesto dreams about a fast transformation of his supply chain and procurement team as in home makeover reality shows. In these TV shows, the team works on the house transformation to ensure that the foundations are solid and builds up on such foundations.

When the team finalizes the work, the house is more functional and has the flexibility to accommodate future needs; for example, more parking space for when the children can drive. When the team reveals the transformed house, the homeowners show their joy in hugs and tears.

As in the home makeovers, in the S.C.O.P.E. of Work, a critical component is the Work that the team does. There is a roadmap or blueprint with frameworks—again that word—processes and tools to go from the original house or situation that no longer fit the needs of Alex's Snacks to the new improved conditions.

It is a full scalable system with inputs and predictable outcomes. As with the home makeovers, too, Maria wants to get the team's tears of joy and a celebration with potato and tortilla chips as it should be!

Maria jumps at the opportunity to help Ernesto's supply chain and procurement team. As with her other adventures—including the Walls at the chocolate factory in Commack, New York—Maria expects a rollercoaster of emotions but has a dogged determination to provide the team at Alex's Snacks with proven and actionable systems, tools, frameworks and processes for greater profits and revenues, lower costs, improved cash flows and asset utilization, without the overwhelm and stress.

Home makeover reality shows present a clear cut between the before and after of the house. The "after" is immediate to the work the team performs. It would be interesting to take a peek in 6 months or a year as a test of the efforts.

Maria goes after sustainable results for Alex's Snacks that the team can keep, replicate, and scale on their own; as NOT seen on TV." All change is hard at first, messy in the middle and gorgeous at the end," as Robin Sharma puts it.

Maria gets her small bag with pink dots to pack, a kind of Super Nanny but for supply chain and procurement. She gets her laptop ready, markers, sticky notes, and other office supplies for her time at Alex's Snacks. Maria hits I-95 early in the morning to be on the way to her next adventure that is about to start.

Endnotes

1. The Trucker News Staff, ATA: *Truck driver shortage remains at near-record high*, The Trucker, 2022, https://www.thetrucker.com/trucking-news/the-nation/ata-truck-driver-shortage-remains-at-near-record-high.
2. Kate Magill, *Nike looks to clear dated products as inventory costs rise*, Supply Chain Dive, 2022, https://www.supplychaindive.com/news/nike-clearing-excess-inventory-holiday-season/633138/.
3. Matt Turner, Jordan Parker Erb, Ben Tobin, and Lisa Ryan, *How Walmart is Trying to Solve its Excessive-inventory Issues*, Business Insider, 2022, https://www.businessinsider.com/walmart-memo-inventory-price-drop-profit-summer-inflation-2022-7.
4. NielsenIQ, *Bursting with new products, there's never been a better time for breakthrough innovation*, NielsenIQ, 2019, https://nielseniq.com/global/en/insights/analysis/2019/bursting-with-new-products-theres-never-been-a-better-time-for-breakthrough-innovation/.

CHAPTER 2:

The Powerful Impact of Supply Chain on the Financials

Key Takeaways

In this chapter, you will learn the following:
- How to read the financial statements
- Know the key elements and how the supply chain impacts them
- How to calculate such impact of Supply Chain/Procurement Actions
- Application examples

Maria arrives at the office building of Alex's Snacks at 7:30 am. In her walk from the parking lot to the main entrance, she thinks about the challenges that Ernesto shared and how to address them. After calling Ernesto to announce Maria's check-in, the security guard lets her in. Ernesto is waiting for her.

"Good morning! Nice to meet you in person," says Ernesto.

"The pleasure is all mine, Ernesto," says Maria.

"Here is the conference room all set up for the training," says Ernesto while opening the door for Maria, "The team will be getting in a few minutes."

"Amazing! I am looking forward to meeting them," says Maria.

"I'm sure this is going to be a success. I won't be able to attend all the training but part of it. You can always reach me if you need help through Teams or text me," says Ernesto.

"Of course, Ernesto! Will do," says Maria.

The training with the five supply chain and procurement team members starts at 8 am sharp.

"Wait! Financials? We don't do that. We're in supply chain," states Jon as soon as Maria shows the first slide of her presentation.

"Thank you!" replies Maria.

Jon opens his eyes as wide as humanly possible. Maria continues to explain that we are following Simon Sinek's approach to the training. In his book Start with Why, Sinek indicates "People don't buy what you do; they buy why you do it. And what you do simply proves what you believe."

Maria starts with the impact of supply chain on the financials because numbers are the language of business. "If you are going to talk business, you need to talk numbers. Avoiding this, it is like coming here to the US, as I did over two decades ago and not learning English. It would be foolish."

Successful entrepreneurs and leaders highlight the importance of knowing your numbers. An example is the multi-billionaire Tillman Fertitta, who is the owner of the Houston Rockets, the Golden Nugget Casinos and Landry's, a Texas-based restaurant and entertainment company. In an interview with Tom Bilyeu, the billionaire affirms that knowing your numbers is a top business principle.

After the why, Maria moves into the how. In her training, Maria considers the three basic types of learners that Jim Kwik, a world-renowned brain expert, categorizes as follows:
- Visual
- Auditory
- Kinesthetic

Each type has a predominant way of learning. Visual learners are those who focus on pictures and images. Auditory learners prioritize sound, like expert talks or podcasts; kinesthetic learners lean more on doing. For every learner type, execution or taking action is critical. As Michael Jordan says: "Some people want it to happen, some wish it would happen, others make it happen." The supply chain and procurement team at Alex's Snacks is in the latter category.

Financial statements: Profit and Loss Statement (P&L), Balance Sheet, and Cash Flows

The numbers of a business are in the financial statements: income statement or profit and loss statement (P&L), balance sheet, and cash flows. As the name suggests, the P&L shows whether a company is profitable; this means that the revenues and gains are higher than the costs and expenses in a period of time. The balance sheet is a snapshot of a company's situation that goes beyond the earnings in a period. It shows assets, debt, and equity. The cash flow statement focuses right on the money and the ins and outs of cash.

Although the supply chain and procurement team doesn't prepare or read the complete financial statements, the concepts and criteria are fundamental for their decision-making and assessing their actions' impact on the business. And because numbers are the language of the business, Maria wants the team to develop the gift of tongues. As the P&L, balance sheet, and cash flow statements serve different purposes, the way to build them is also different.

Cash Basis vs Accrual Accounting

For the financial statements, there are two types of foundations or stepping stones or bricks—if you love Legos like Maria's son: 1. cash basis or 2. accrual. Cash basis is straightforward. Accrual can be tricky.

Cash basis is "Show me the money" in Jerry Maguire, a movie with Tom Cruise. It's Benjamin Franklin's face. It's 100-dollar bills. It's 20-dollar bills. It's the money the company gets and the money the company pays. When

Alex's Snacks sells the jalapeño potato chips to a consumer that buys online, that's a sale in cash. It's money that the company is getting. When Alex's Snacks pays the corrugated box supplier, it's money out. Simple as that.

Accrual accounting considers revenues and expenses when they occur, independent of payments. When Alex's Snacks delivers thousands of hot tortilla chips to Walmart, Walmart pays months later. But the moment to record this sale is with the delivery of the goods. When Alex's Snacks receives roll stock film, the company pays the supplier in 60 or 90 days. Under accrual accounting, the company records the transaction with the goods receipt. It doesn't wait until payment.

The main difference between cash basis and accrual accounting is the timing to recognize and record the transactions. Cash basis—"Show me the money"—tracks money in and out, while accrual accounting traces receipts and deliveries of products to anticipate payments.

At college, Maria battled against accrual accounting. As Rhett Butler said in Gone with the Wind, "Frankly, my dear, I don't give a damn." She didn't give a damn back then because Maria considered it too theoretical, a concept as up in the air as soap bubbles. The struggle was bringing that bubbly concept down to Earth. Maria spots the same feeling in Sarah when in a low tone of voice and breaking up she asks, "I have heard Finance talking about accruals. Why would they use this concept as opposed to cash flows? Isn't it money that matters?"

Maria shares with Sarah that in the past, she had a similar extraterrestrial encounter with the accruals, but she realized that Finance uses both cash basis and accrual accounting. It is the best of both worlds, as with carrot cake, when you eat your veggies too, as Elmo from Sesame Street or Maria's husband would say.

The cash flow statement follows cash basis—no catch here—the "show me the money" approach. It shows cash in and out over a period of time. With a projected cash flow statement, a company can identify any gaps to cover or the need for additional funds.

Then there's the balance sheet and P&L. "Any guesses?" Maria asks the team members to keep them awake or wake them. Both balance sheet and P&L follow accrual accounting to anticipate payments as a baby shower anticipates a newborn, a wedding shower a marriage, and an RFP (Request for Proposal) anticipates cost savings in the procurement world.

Following accrual accounting, companies realize the RFP cost savings with the goods receipts at the new lower price, no matter the timing for payment. Alex's Snacks makes the payment to the supplier in 60 days (at some time in the future), but the new lower cost hit the P&L before; hurray! Price increases from suppliers—thumbs down, boo from procurement—also hit the P&L with the good receipts and not at the moment of payment. Before it gets salty and spicy with the key elements of the financial statements, Maria challenges the team with a few questions.

Chips for Thought

1. Per her conversation with Ernesto, there is excess inventory for some items. As Alex's Snacks doesn't need to buy more, the company doesn't issue new purchase orders. Can the supply chain and procurement team report cost savings for the P&L?
2. The team at Alex's Snacks creates a PO at a price 25% lower. Does this cost reduction hit the P&L now?
3. There is a delay with a supplier's order for oil. The requested delivery date is today, but the revised delivery date is in a month. The good news is that this order has a 10% lower unit cost. Does this benefit reflect in today's P&L?

After the short quiz, "Let's double dip into the financial statements," suggests Maria.

Cash Flow Statement

The cash flow statement shows the inflows and outflows of funds—the ins and outs of the chips. It covers a period of time. Examples are a year, a quarter, a month, etc. A cash flow statement can be historical or forward-looking, a projection or an estimate.

There are three distinct sections:
1. *Operating activities*—they are the business's bread and butter or the tortilla chips and salsa verde. They are the regular business operations that generate revenue and expenses.
2. *Investing activities*—they are not normal (an adjective that becomes popular in Covid times) or part of the regular operations of a business. For example, selling machines or equipment.
3. *Financing activities*—these activities include both debt and equity for the business to support operations.

There are two ways or recipes to build the cash flow statement:
1. *Direct method*—This is to start from scratch. The person preparing the cash flow statement gathers all the transactions for a specific period and categorizes the transactions into 3 buckets: operating, investing, and financing activities.
2. *Indirect method*—Without getting into details that only accountants find charming, the person building the cash flow statement starts with the P&L or income statement and makes some adjustments without the need to start from the ground up.

"Why the heck are we doing this?" This question escapes Mike's lips; he's the most experienced team member. "I'm sorry. I don't get it," Mike justifies, with the pungent taste of sour cream in his expression.

"No need to apologize. It is a great question that I would like to answer with an example of cost savings, if I may," replies Maria. Mike nods his head.

Maria knows that supply chain and procurement teams chase cost savings like Tom the cat does it with Jerry the mouse or Sylvester the cat with Tweety bird. When the team captures them—after the non-stop chasing—the heavens open and the trumpets play. This Hallelujah moment ends when finance sees the cost reduction as delusional. This hurts as much as having all your wisdom teeth removed at once without anesthesia. It. Hurts. All the hard work, sweat and tears, for finance and other teams consider procurement as a bunch of liars! In their heads, procurement can't be trusted.

This happens because procurement, in presenting the savings to finance, doesn't include a projected cash flow with the purchases. Oops! The typical approach consists of showing current cost, standard cost, or average cost against the new lower cost. This is good but not enough. Finance also needs to see how much the company will be paying in the next 12 months to compare against the current situation or baseline. Show me the money, baby!

P&L

In addition to the cash flow statement, Maria recommends that the supply chain and procurement team presents the projected P&L in their conversations or fights with finance. A P&L can be historical or forward-looking, a projection or an estimate. The nickname of the projected P&L is the Halloween word "budget."

Both the P&L and cash flow statements are fundamental. Showing only one is like eating mac and cheese without either the macaroni or the cheese, a peanut butter and jelly sandwich without the peanut butter or the jelly. Maria takes the opportunity for a quick recap by asking the team members about the goals and criteria of the P&L. Tim, a recent college graduate from Michigan State University, jumps out of his chair to answer the question.

"The P&L shows income and expenses for a period of time and follows the accrual accounting method," says Tim, hitting the jackpot. The profits (or loss) are in the P&L. The P&L shows the performance of a company. It has a similar structure as that of the cash flow statement—operating, financial, and tax activities—with the following components:

+ Revenue
- Cost of goods sold (COGS)

Gross profits
- Operating expenses

Operating profit
- Non-operating expenses (excluding depreciation and amortization)

EBITDA (earnings before interest, depreciation, and amortization)
- Depreciation
- Amortization

EBIT (earnings before interest and taxes)
- Interest and Taxes

Net profit

Supply chain and procurement have a huge impact on revenue, COGS, gross profits, and all operating expenses. This impact is as gigantic as blue whales, colossal squids, and African elephants together.

As the P&L follows accrual accounting, finance records the transactions based on receipts and deliveries, not payments or money ownership exchanges. "Let's see in action how powerful you are," Maria says to the team members.

Revenue

This is one of Maria's favorites, the top line of the P&L. Traditionally, supply chain is a synonym for cost containment and optimization. While savings are a top priority, supply chain goes beyond cost reduction.

Like its competition, Alex's Snacks is selling online on Amazon and through its e-commerce store. With almost 500 billion in sales and almost 1.5 million employees, Amazon dominates the e-commerce world.

Jeff Bezos, former CEO, indicates that the key to success is a relentless customer focus. Amazon strives for happy customers. Customers are happy when they have a great assortment to choose from, get their orders delivered on time, and get support from customer care. In other words, customers are happy because of Amazon's supply chain.

Supply chain plays an essential role as a catalyzer for growth, orchestrating the multiple moving parts from raw material acquisition until delivery to the customer we want to delight. Not considering the impact of supply chain on revenue is leaving money on the table.

Marketing may sell the sun, the moon, and the stars with delivery in the next 24 hours, but with no product available, there are no sales; instead, there are disappointed and angry customers, like angry birds kicking and screaming for worms. Nasty. And how about the negotiated savings from procurement? "I don't care," as Ed Sheeran and Justin Bieber sing. There is nothing to sell. No sale. No gain. Pure cost-based, traditional supply chain concept, hasta la vista!

Cost of Goods Sold

All roads lead to Rome, so as to COGS. In the ancient Roman empire, the road system had the center in the capital, with all paths connected to Rome. "The impact that you guys with your activities in supply chain and procurement have on COGS is humongous. There are times that you may not realize the impact or the extent of it," Maria says.

COGS is all costs related to getting the snacks produced and ready to sell to the consumers eagerly waiting to devour them. It includes the ingredients such as corn, oil, sea salt and cheddar cheese, the packaging, freight-in cost, plant labor, and factory overhead. Per Ernesto's explanations, Alex's Snacks works with co-manufacturers. That cost is part of COGS, as all production costs of the crunchy and savory chips.

Production or factory costs are split into **direct** and **indirect**. Direct costs are those connected with a specific product. An example of a direct cost is when the line runs to produce barbecue chips, as that cost can be assigned to these spicy sweet chips. An example of indirect cost is the salary of factory supervisors because they can't be related to a particular product. Another example is the factory utilities that are included in the overhead cost.

Production or factory costs also split into **variable** and **fixed** costs. Variable costs increase or decrease in the same proportion as production does. If

Alex's Snacks doubles production of their original potato chips, the cost of ingredients, packaging, and assembly-line worker wages will double too.

With production changes, fixed costs do not alter as the pines, live oaks, palms, and cedar trees do not change leaves with the seasons. For instance, increasing to two times the original potato chips doesn't have an effect on factory supervisor cost. Her salary remains the same, so there is no change in cost. There are some situations in which fixed costs change, but not in the same ratio as production. In the example, the plant supervisor may need to work overtime.

Maria senses that the team is feeling like getting out of Six Flags Magic Mountain Full Throttle, the number one tallest and fastest looping roller coaster in the world. With his comment, Ruben, a new team member, says a big yes to Maria's intuition: "I need to go for a tequila after all these costs!" A break with snacks follows for the team members to relax and regain energy.

Once the team members are back, Maria reviews COGS and throws more Chips for Thought at them.

Chips for Thought

1. Are supply chain planners' and buyers' salaries part of COGS?
2. How about the cost of receiving materials for production at the co-manufacturers?
3. If Alex's Snacks needs to apply stickers to the packaging because of an error with the artwork, is this labor cost part of COGS?
4. Are the generic corrugated boxes or shippers direct or indirect costs?
5. Is the plate cost for new artwork variable or fixed?

Gross Profits

If Ernesto had the genie's magic oil lamp like Aladdin, higher gross profits would be one of his wishes. What executive would ask for something different? Gross profit is the difference between revenues and COGS. Companies also calculate the gross margin, the percentage of gross profit over revenue (gross profit divided by revenue).

Gross profit is not what the company gets. There are other costs that Alex's Snacks faces to produce their potato and tortilla chips. "Remember that COGS cover production costs only," Maria emphasizes.

Operating Expenses

Operating expenses (OPEX) include the costs to support operations such as general & administrative, marketing, rent, utilities, and services. Indirect purchasing or indirect procurement has a profound impact on OPEX when they negotiate service contracts on technology—examples are SAP S4 Hana or JD Edwards—and professional services, among others.

Operating Profit or Operating Income

Drumroll, trumpet, trombone, saxophone and ukulele ...ladies and gentlemen of the supply chain and procurement team...this is the introduction to operating profit, the forever and ever Hollywood star of business.

The operating profit is the net earnings from the core business, from the chips. The focus is on performance. Its calculation is as follows:

Gross Profits = Revenue - COGS

Gross Margin (%) = Gross Profit/Revenue

A positive operating profit means that the company is performing. A negative result from operations means that costs are higher than revenues and this loss is a biggie for management.

Operating Profit/ Operating Income Calculation

+Revenue
- COGS
- OPEX
- Depreciaton
- Amortization

Operating Profit/ Operating Income

"Is operating profit the same as EBIT and EBITDA? Private equity companies use these acronyms but not sure if I'm totally off with this" says Amy from the team.

"Thank you for bringing this up," Maria replies. "You're right. There's confusion and misunderstanding around these three different terms" Maria adds.

EBIT
E – earnings
B – before
I – interest
T – taxes

EBIT is not the same as operating profit or operating income. While both consider revenues, COGS, OPEX, depreciation and amortization, EBIT also includes non-operating or other income like that coming from bonds and stocks.

Other Income

The income listed in this line item of the income statement included in the EBIT calculation does not come from its core business of chips. For instance, the proceeds from selling a potato chip packing machine are other income. Selling this kind of machine is not Alex's Snacks main activity; selling chips is.

EBITDA

E - earnings
B – before
I – interest
T – taxes
D – depreciation
A – amortization

EBITDA is different from both operating profit or operating income and EBIT. "I think I know the difference between EBITDA and EBIT: the D and the A," jokes Ruben.

"Correct, Ruben! EBITDA doesn't include depreciation and amortization," says Maria.

The objective with operating profit, EBITDA, and EBIT is to show the results of the business without considering the financial aspects—capital and debt—to support operations or the taxes to Uncle Sam.

"As Ruben noted, EBITDA doesn't include depreciation and amortization," Maria states. Depreciation is the loss of value of a fixed asset (machine, computers, etc.) over its useful life because of wear and tear. Amortization is the same concept applied to intangible assets like goodwill, patents, and trademarks.

On the whiteboard, Maria writes the following:

+	Revenues	+	Revenues	+	Revenues
-	COGS	-	COGS	-	COGS
=	Gross Profit	=	Gross Profit	=	Gross Profit
-	OPEX	-	OPEX	-	OPEX
-	Depreciation	-	Depreciation	-	Depreciation
-	Amortization	-	Amortization	-	Amortization
=	**Operating Income/ Operating Profit**	=	**Operating Income/ Operating Profit**	=	**Operating Income/ Operating Profit**
		+	Non-Operational Net Income	+	Non-Operational Net Income
		=	**EBIT**	=	**EBIT**
				+	Depreciation
				+	Amortization
				=	**EBITDA**

Interest and Taxes

A company finances its operations with a combination of debt and equity. This combination is the capital structure. When the company uses debt like bond issues or loans, the P&L shows interest expenses. Alex's Snacks is financing part of its growth with debt.

Before Alex's Snacks can take the earnings after considering the interest expense, there are expenses for federal and state income taxes (Yeah! A portion goes to the IRS, friends).

Net Profit/ Net Income Calculation

+ Revenue
− COGS
− OPEX
− Depreciaton
− Amortization
+ Non-operating income
− Interest, non-operating expenses
− Taxes

Net Profit/ Net Income

Net profits—Bottom line

The net profits or bottom line is the finished line. The moment of truth. The "To be, or not to be" of William Shakespeare's play, Hamlet. Net profits ARE what the business makes after all costs, including taxes. If the result is positive, Alex's Snacks can distribute or reinvest. A negative result means that the company is not profitable; its costs are higher than its income.

Maria and the team get ready for more fun with the balance sheet. This is the last core financial statement to cover.

Balance Sheet

The balance sheet is like a snapshot—a Kodak moment—of a business's finances. The balance sheet can be historical or forward-looking, a projection or an estimate.

These are the three big chunks (or chips).
1. Assets
2. Liabilities
3. Shareholders' equity

The balance sheet gets its name because of the fundamental equation:

$$\text{Assets} = \text{Liabilities} + \text{Equity}$$

It is the equilibrium between assets on the left side of the seesaw and liabilities and shareholders' equity on the right side. Liabilities and shareholders' equity finance assets.

"Now that we know what we can find on the balance sheet, let's dig deeper and savor the assets, liabilities and equity," says Maria, with intention in her voice.

"Whoa, Maria, it looks like it's gonna be intense," says Jon.

"We may choke," adds Ruben.

"Agreed," says Sarah.

"Guys, it's gonna be fine," Mike calms the team down. Maria proceeds with the explanations.

Assets

Assets are resources that a business owns or controls. The business expects assets to provide current and future benefits, to generate sales.

There are two pockets: 1. current assets and 2. non-current assets

The most liquid assets go first in each of these pockets—current and non-current assets.

Current means that the assets can turn into cash **sooner than in 12 months**. That is the stop in time, as the magic with Cinderella is at midnight. Current assets include:
1. cash and cash equivalents
2. marketable securities
3. accounts receivable (AR)
4. inventory
5. supplier prepayments
6. prepaid expenses

"All right, you now get my attention. We have so many issues with inventory. It is…it is…a mess," says Jon.

"Most companies face issues with inventory. At college, we saw different case studies. With challenges come opportunities," says Tim.

"And headaches," adds Mike, continuing, "Since I am here, that is about 9 years, the inventory stuff has been bad, and with Covid, we bought like crazy, and now we have piles of inventory."

"Inventory is a hot topic, like the oil when frying the chips. As such, we will devote time to optimizing its levels. Now, let's get back to our assets in the balance sheet with a brief explanation of the most common ones," Maria says.

Current assets include:
1. Cash and cash equivalents: These are the most liquid of all assets. Available. Ready to use.
2. Marketable securities: A company can convert these assets into cash with short notice. They have a maturity of three months or less.
3. Accounts receivable (AR): Sales revenue on credit. When Alex's Snacks sells to Sam's Club, the company doesn't get paid until months later, after the delivery of the goods. The amount owed is in AR. When Alex's Snacks gets the payment, it reduces AR and increases cash and cash equivalents for the same amount.
4. Inventory: Top priority for the team. There is a whole section on it!
5. Supplier prepayments: These are payments in advance to suppliers before receiving the materials or products.
6. Prepaid expenses: The company has made the payment in advance; for example, marketing campaigns and insurance.

Non-current means that the assets can turn into cash *after 12 months*. This is like Cinderella after the clock shows 12 am when the horses turn back into mice and the carriage into a pumpkin.

Non-current assets include:
1. Fixed assets like property, plant, and equipment (PP&E).
2. Intangible assets such as intellectual property and goodwill.

"These are the assets on the left side of the equation. Let's move on to the right side. What do we have there?" Maria asks.

Amy says, "Per my notes, liabilities and equity. They finance the assets."

Touchdown for the team!

Liabilities
A liability is money that the company owes to third parties, including suppliers, creditors, Uncle Sam, and employees. Liabilities also show on the

balance sheet based on their liquidity. As with assets, there are current and non-current liabilities. Current liabilities are the amount due in 12 months, while non-current liabilities are the amount due after 12 months.

"Thinking about my mortgage here. What happens with a loan over 20 or 30 years on which we have monthly payments? Where does it go?" asks Amy.

After pausing, Sarah answers, "Part in current and part in non-current liabilities."

"Excellent! Way to go!" Maria encourages the team.

Current liabilities include:
1. Accounts payable: the opposite of accounts receivable; as the sun and moon, as profit and loss. Alex's Snacks pays their suppliers on average in 60 days. The amount that the company owes is in AP. Likewise, when Alex's Snacks makes a payment, it reduces AP and reduces cash and cash equivalents for the same amount.
2. Wages payable: these are liabilities for wages earned but not yet paid. As such, the amount in wages payable is a short-term obligation, due within 12 months.
3. Interest payable: this is the amount of interest owed.
4. Dividends payable: this is the amount of dividends owed. This can happen when the company approves the dividend but needs to make the payment.
5. Customer prepayments: "The opposite of," Maria starts saying, when Tim, like the compressed cork in a champagne bottle pops up with "supplier prepayments."
   "You've got this, team!" Maria celebrates.
6. Current portion of long-term debt: this is the situation that Amy brought up referring to her mortgage.

Non-current liabilities include:
1. Bonds payable: companies may issue bonds to get funds to finance their operations. This is the amortized (remaining) amount of the bonds.
2. Long-term debt: the amount due after 12 months, based on the debt schedule. The debt schedule shows the outstanding debt, the interest

expense, and the payments against the borrowed capital that the company needs to make in every period.

Equity

Equity—also known as shareholder's equity or net worth—consists of what the owners or shareholders own. The equity calculation is as follows:

Equity = Total assets - Total liabilities

Equity has two main items:
- Paid-in capital—the dollar amount that the owners paid when the company started or, in the case of the shareholders, the dollar amount paid with the first stock issue.
- Retained earnings—the profits kept at the company for reinvestments.

Paid-in Capital

This amount can be small, like the incredible story that Warren Buffet tells about "Mrs. B" Blumkin and her furniture store[1] that went from almost no capital to millions.

Mrs. B started the business with a $500 loan from her brother and the proceeds from selling used clothes with her husband.

"Quick question," Maria says, "do we know the equity value at that moment?"

"No, we don't because we don't know how much money she got from selling clothes. All we know is that the initial investment was $500 plus," says Mike.

"We don't, but the $500 shouldn't be included because it's a loan," corrects Tim.

"Who agrees with Mike? Who agrees with Tim? What do you think?" Maria asks to toss seasoning on the discussion.

The correct answer is Tim's. The $500 is the amount that the furniture store of Mrs. B owed to an outside party (Mrs. B's brother). The paid-in capital was the amount from selling second-hand clothes.

The paid-in capital can be large too. There can be common stock and preferred stock. There can also be additional paid-in capital or capital surplus. This is the amount that the shareholders have invested above that of common and preferred stock.

Retained Earnings
This is the generated profits or net income that the company does not distribute. It is the umbilical cord with the P&L. The P&L shows the net profit on the bottom line. From there, the company may decide to distribute. That part goes to the pockets of the shareholders. What is left goes to retained earnings in the balance sheet.

"Phew, the three core financial statements covered! Let's now jump into the action," Maria invites the team.

Five FAQ—Frequently Avoided Questions—on Supply Chain and Procurement Impact on the Financials

This is where the rubber meets the road, where Alex's Snacks meets the challenges, where the procurement and supply chain team meets the financial world. When a company is growing and expanding, like the Roman Empire in the 6th century BC or the Inca civilization in Peru in the early 13th century, teams need to be prepared and equipped to make the best decisions.

Alex's Snacks is growing with its own brands and acquiring others in a ridiculously short amount of time, like Alexander the Great, a Macedonian king did, when conquering Western Asia and Egypt in a few years. Alex's Snacks is not in a war, but kind of.

The procurement and supply chain team battles with challenges and decisions. They need to understand the impact of their actions on the financial statements. This is their armor to protect the company. Ernesto cannot be involved in everything.

"Here are seven common decision challenges that I've found that supply chain and procurement teams face when the business is growing exponentially. Fast. Superfast. Take some chips, get ready and buckle up," advises Maria.

Purchasing In Larger Quantities

With more sales, manufacturing companies buy more ingredients and packaging. E-commerce businesses buy more turnkey products to sell.

With an increased volume, suppliers offer quantity discounts, and like McDonalds' popular slogan, "I'm lovin' it," procurement loves these attractive discounts. The team thinks they are taking advantage of the new leveraging power and feel good, so good, so good, I got you, as James Brown sings.

"Enticing as they may seem, discounted pricing based on quantity breaks doesn't always turn into a lower COGS or OPEX in the P&L," Maria says.

"What?! Hold on. How's that even possible?" argues Jon.

"Leverage volume, leverage volume is what we are asked to do, and it makes sense, don't you think?" Jon asks while looking at the other team members with defiant eyes.

"Let me show you something," answers Maria while Sarah, Mike, Tim, Ruben, and Amy remain silent.

Maria shows the gross margin simulator for finished goods—on the resources page at https://usmsupplychain.com/resources/—as it is simpler to explain than that for raw materials and packaging. Besides, in the training there is a whole module on inventory.

Finished Goods Margin Simulator	
Top Line Information	
Projected Sales – Qty	300,000
Average Price	$ 3.0000
Total Projected Sales	$ 900,000
Estimated Lost Sales	$ 0
Total Sales - Lost Sales	$ 900,000
Inventory	
Qty Needed	500,000
MOQ (minimum order quantity)	600,000
Qty to Buy	600,000
Unit Cost	$0.5500
Purchase Cost	$330,000
Order Processing Cost	$1,200
Financial and Warehousing	
Capital	$23,100
Storage	$16,500
Maintenance	$9,900
Damaged and/or obsolete materials	
Disposal	$3,300
Other	
Shipping	
Shipping Cost	$2,000
Expedited Shipping	$2,800
Late Delivery Penalties	$400
Total Cost	$389,200
Gross Margin	$510,800
Gross Margin %	56.76%

The gross margin simulator has the following sections:
1. **Top line information:** Supply chain has an impact on the top line of the P&L. When Alex's Snacks doesn't have the products ready on time, the company cannot sell. This section of the simulator considers such lost sales.
2. **Inventory:** This section considers the minimum order quantity (MOQ). This applies to new product launches more than when the company buys in large volumes. The simulator compares the quantity to buy against the MOQ and takes the greater quantity for the calculations. This section also accounts for the cost of processing a purchase order. There are hands, brains, and a system when creating a PO.
3. **Financial and Warehousing:** "This is it," Maria addresses Jon, like the "Eureka!" from Archimedes when he discovered the principle of buoyancy. "We have the invisible costs exposed," says Maria, like when the media shoots photos of famous artists without makeup or revealing those extra pounds.

This section of the simulator sheds light on the following costs:
- Capital
- Storage
- Maintenance
- Damaged and/or obsolete materials
- Disposal

"Nice, but I don't get it," says Jon.

"How the heck are we gonna use this?" Mike supports Jon in his argument.

"I understand your concerns, Mike and Jon, but the simulator is powerful and simple to use," Maria assures them.

"When you buy in larger quantities, you have more capital tied up that you can't invest somewhere else and also need space to place the inventory... not to mention that it can become aged and obsolete," explains Maria.

"And that makes the savings from the quantity discount evaporate, correct?" asks Amy; Maria nods.

"Got it, Ms. Maria," adds Ruben.

"In any case, it's gonna be hard to come up with those costs," says Jon, as if waving a white flag in this lost fight, but with honor.

4. **Shipping**: The gross margin simulator has this section that includes the shipping cost, expedited shipping, and late delivery penalties.

"We can make it on time but at a higher cost with expediting. We don't have another option in many circumstances, though," adds Sarah.

"And those chargebacks from Walmart kill us, so we need to eat the cost of expediting. We are between a rock and a hard place," says Mike, frowning.

Maria adds that the gross margin simulator can help the team with more. All are anxious to see, like kids waiting for the magician to get a rabbit or a pigeon out of a big box.

Not Having the Material On Time

Another functionality is quantifying lost sales. Not having the material on time leads to lost sales. Lost sales are as intricate as the soap opera "The Bold and the Beautiful" or a big corn maze.

"Where do lost sales go in the financial statements?" asks Maria.

"I think that we didn't cover that," answers Amy, flipping the pages of her notebook.

"You are right, Amy," confirms Maria, "lost sales are not in the P&L or income statement, not in the balance sheet, and not in the cash flows."

"So what do we do with them?" asks Jon, opening his arms like Christ the Redeemer in Rio de Janeiro, Brazil.

"We calculate lost sales to help us make decisions. They don't affect the financial statements because they are an estimate. They are not 100% real," explains Maria.

"I'm confused," claims Sarah, "...and the accruals? Are they real?"

"No, they're ghosts...boo! boo!" teases Ruben.

Jokes aside, Sarah asks an excellent question. Accruals are estimates too, but the loss goes into the P&L, and the accumulated accruals go to the balance sheet. Lost sales are estimates, and they don't go anywhere.

This happens because of the prudence or conservatism accounting principle included in GAAPs (Generally Accepted Accounting Principles). This principle states that a company should recognize income when it is realized (delivery of the goods); when there is 100% certainty. The same principle states that a company should recognize losses as soon as possible, even when there is uncertainty about the outcome.

As a summary, Maria says, "The rule of the game is to play safe. We recognize revenues when we are 100% certain about the delivery of the goods while we recognize potential losses immediately, even if we are 1% certain about the potential loss."

"Oh my gosh, why do we need to know this accounting stuff if we are not accountants? We're in supply chain," says Jon, placing both hands on top of his head.

"Really, Jon? You don't want to be an accountant?" Maria smiles, "Just kidding!"

She continues, "I talk about this because it is important to consider lost sales in making decisions and knowing how the heads in finance work. The same prudence principle applies to cost avoidance. For finance, cost avoidance is like smoke and mirrors because it doesn't go to the financial statements. When presenting solutions and making recommendations, we need to be mindful of this."

Shipping in Larger Quantities

Shipping a full truck load (FTL) or a full container load (FCL) makes the unit cost lower compared to shipments of less volume. There are other advantages. When the company fills up the truck or the 40' or 20' container, the lead time is shorter, and the risk for damage is lower than with a consolidated shipment.

The gross margin simulator considers the warehousing cost and other unpleasant surprise costs for the team's decision-making. For instance, Alex's Snacks has bitter and salty experiences with its co-manufacturers rejecting the loads because of not enough space, re-directing shipments and paying for temporary storage. It can become messy and costly if all these factors are not weighed in.

Payment Terms Extension to Suppliers

When a company scales up in sales volume, a common practice is extending terms to their suppliers based on the new increased volume. More sales mean more purchases. When Alex's Snacks started operations, they didn't have terms with suppliers as the business was paying on receipt. Then it moved to Net 30, Net 45 and now most payment terms average Net 60 days. Procurement plans to extend them to Net 90 days.

When this happens, suppliers may increase their prices to compensate for the capital cost to them that applies to getting paid in an extended period of time. Maria has seen that supply chain and procurement teams do not have a good way to understand whether such price increases offset the benefits of extending terms.

"Which financial statements does a payment term extension to suppliers impact?" asks Maria, and adds, "I am giving away these crunchy tortilla chips for the correct answer."

"Payments sound related to cash flow statement," says Tim.

"Here you go, Tim," says Maria, handing a 4.4oz tortilla chip bag to him.

"Any other financial statements that could take the heat?" asks Maria indicating there is another answer.

"P&L!" say Amy and Sarah at the same time, as if they're answering a question on Jeopardy!

"Correct, ladies!" says Maria, handing them 4.4oz tortilla chips bags to them too. "If the supplier increases their prices or charges interest, that added cost for Alex's Snacks goes into COGS."

"Another question for the gentlemen over there, Jon and Mike," says Maria, keeping the momentum going. "How do you know when extending the supplier terms is beneficial?"

Mike answers, "I will give it a shot."

"Hope it is not of tequila," says Ruben with his usual sense of humor.

"I think it will depend on the cost of capital because, you know, the company can invest or use the funds to get a return. If the return is higher than the supplier's price increase, it's a win" states Mike. "What do you think, Jon?"

"Yeah, I guess if the return for investing the funds is higher than the supplier's increase, I'd go for it."

"Excellent job! More tortilla chips to celebrate," says Maria, giving away four 4.4oz bags. "Jon, Mike, four to double the fun! Amazing work, everybody!" she adds.

Liabilities and Accruals

"This is your topic!" exclaims Maria, looking at Sarah to introduce liabilities and accruals.

"Can't wait," Sarah agrees.

Buying more to support the increased demand goes hand in hand with more POs, with more commitments with suppliers. Previously, in the training in a "Chips for Thought," Maria explained that when the team creates POs, they don't show in the financial statements because there are no receipts of goods. This is scary for Finance, like Freddy Krueger in the Nightmare on Elm Street horror film series.

Let it sink in: more commitments with suppliers that imply future payments, and they are not in the financial statements. This is one of the reasons for accruals. Finance creates a reserve (accrual) to be prepared, like Scouts, to face the payment obligations when they come. This reserve is a loss in the P&L and a liability in the balance sheet. With the payments to suppliers, the amount in reserve goes down.

Another reason for accruals pertains to inventory hold at the co-manufacturers. When the co-manufacturers operate under the contract manufacturing model, they buy the ingredients and packaging to support Alex's Snacks operations. The co-manufacturers own the inventory, so it is not on the company's books. But Alex's Snacks is actually liable for that inventory. To show this on the books and avoid a "Frankenstein's surprise," finance may accrue for the off-balance sheet's inventory.

Endnotes

1   Richard Feloni, *Why Warren Buffett Considers the Deal He Made With an 89-year-old Woman one of the Best of his Career*, Yahoo, 2015, https://finance.yahoo.com/news/why-warren-buffett-considers-deal-143539508.html.

CHAPTER 3:

Metrics

Key Takeaways
In this chapter, you will learn the following:
- Financial ratios
- Benchmarking with other CPG companies
- Define key metrics in supply chain in alignment with finance and strategy
- Build a strategic goal tree for execution
- SCOR framework revised in 2022

Finance-Supply Chain Alignment on Key Metrics

The language of business is numbers, and the numbers speak for the business. There are five main categories of key performance indicators or ratios:

Liquidity

Liquidity ratios help to determine a company's ability to pay its short-term debt obligations without adding external capital. The focus is on the cash flows—like Jerry Maguire yelling, "Show me the money"—and how the business can meet its obligations.

Some examples are the current ratio, quick ratio, and operating cash flow ratio.

"Let's dive in!" says Maria.

Current Ratio

This ratio shows a company's ability to pay off its current liabilities—the amount owed within 12 months—with its current assets like cash, accounts receivable, and inventory. The higher the ratio, the better the liquidity.

$$\text{Current ratio} = \text{Current assets}/\text{Current liabilities}$$

Quick Ratio

"The quick ratio," starts Maria, when Ruben interjects, "It's fast," teases Ruben. The whole team giggles. Maria explains that this ratio—also known as "acid-test"—measures a company's ability to pay its short-term obligations with its most liquid assets.

"Where can you see these most liquid assets?" asks Maria.

"Balance sheet," answers Tim, like answering in Wheel of Fortune to win a prize.

"Perfect!" says Maria, "The most liquid assets are at the top of the balance sheet."

The quick ratio considers cash and cash equivalents, marketable securities, and accounts receivable, leaving inventory out of the formula—and out of the picture too!

$$\text{Quick Ratio} = (C + MS + AR)/CL$$

where
C = cash and cash equivalents
MS = marketable securities
AR = accounts receivable
CL = current liabilities

Another way to calculate the quick ratio is as follows:

$$\text{Quick ratio} = (\text{Current assets} - \text{inventory} - \text{prepaid expenses})/\text{Current liabilities}$$

Metrics

Operating Cash Flow Ratio

This ratio shows the company's ability to pay off its current liabilities—due within a year—with the cash inflow generated by its core business operations.

Operating cash flow ratio = Cash flow from operations/Current liabilities

This formula applies where operating cash flow is the amount of cash that a company produces from its core business operations.

"Where can we see the operating cash flow to plug in the formula?" asks Maria.

"Cash flow statement," answers Amy.

Time to move on to profitability ratios.

Debt or Solvency

Debt or solvency ratios measure a company's ability to meet its long-term debt obligations. They are indicators of the financial health of the company.

"Are these the ratios that banks use for lending money?" asks Amy.

"That's exactly right," answers Maria and adds, "Banks want deals that give good bang for their buck."

"Examples of solvency ratios are Debt-to-Equity Ratio, Equity Ratio, and Debt Ratio," Maria explains.

"Hold on. Didn't we see this already? Isn't the same as liquidity ratios?" asks Mike.

"Trying to find that," says Amy, thumbing through her notebook.

"Yes, we have seen liquidity ratios, but these ratios are different. It's easy to mix them up," admits Maria. "Let me show a table I've found online with a comparison."[1]

Solvency Ratios	Liquidity Ratios
Measures financial health of company	Measures financial health of company
Focuses on long-term stability	Focuses on short-term stability
Includes all assets, such as inventory	Emphasizes cash and cash-like holdings

"I feel much better," says Amy, taking a deep breath.

"Great!" Maria says, "Let's go over the solvency ratios!"

Debt to Equity Ratio
It measures the company's financing that comes from creditors and investors.

$$\text{Debt to Equity Ratio} = \text{Total liabilities}/\text{Total equity}$$

Equity Ratio
The equity ratio indicates the percentage of funds received from the shareholders.

$$\text{Equity Ratio} = \text{Total equity}/\text{Total assets}$$

Debt Ratio
It indicates a company's ability to pay off its liabilities with its assets.

$$\text{Debt Ratio} = \text{Total liabilities}/\text{Total assets}$$

"This bird's-eye view is sufficient for you to have an idea. We focus more on the profitability and efficiency or activity or operational ratios," concludes Maria.

Profitability

Profitability ratios assess a business's ability to generate earnings based on its revenues, operating costs, assets or investment, or shareholders' equity over time.

The company calculates these ratios with the data from a specific moment. Some common types include gross profit ratio, operating profit ratio, net profit ratio, and return on investment (ROI).

"Do any of these terms ring a bell?" Maria asks. "Ring a bell?" I think the bells are ringing in my head so hard that it's gonna explode," says Ruben.

Everyone in the room laughs.

"We've seen gross profit, operating profit, net profit," says Sarah.

"Amazing! The ratio calculations are as follows," details Maria:

Gross Profit %
   Gross Profit % = Gross Profit/Revenue of Operations X 100

Operating Profit %
   Operating Profit % = Operating Profit/Revenue of Operations X 100

Net Profit %
   Net Profit % = Net Profit/Revenue of Operations X 100

"We haven't covered ROI, but I'd bet dollars for donuts that you've heard about it," Maria says.

"How can we not? Mike says, "These days, it's all about ROI and savings."

The Return on Investment (ROI) or Return on Assets (ROA) measures how a business can generate profits from its liabilities and shareholder's equity.

The formula is as follows:

ROA = Earnings before Interest and Taxes (EBIT)/Assets X 100

Operating Performance, Efficiency, or Activity Ratios

These ratios that are expressed in different terms—efficiency, operating performance, and activity—measure how efficient the company is with the use of the resources to generate sales and how well assets convert into cash.

It is a big woo-hoo when a business has great levels of cash inflows and sales with few resources, like the TV show character MacGyver, who achieved huge accomplishments by using only a Swiss Army knife and duct tape.

These ratios provide insights for a company to improve its operational, asset management, and other business aspects.

"Holy cow or holy chips, as Robin would say to Batman; this is the perfect opportunity for you supply chain and procurement people to shine," says Maria, smiling with her eyes.

"Let's go over the key efficiency ratios to measure YOUR impact at Alex's Snacks!" proceeds Maria.

Metrics

Ratios	Liquidity Ratios
Efficiency Ratio	Efficiency ratio = expenses/revenue
Operating Efficiency or Operating Ratio	Operating efficiency or operating ratio = expenses (operating expenses + COGS)/net sales
Accounts Receivable Turnover Ratio	Accounts receivable turnover ratio = net sales/average accounts receivable
Average Collection Period (non-ratio metric related to accounts receivable turnover ratio)	Average collection period = days in period to track/accounts receivable turnover ratio during period
Accounts Payable Turnover Ratio	Accounts payable turnover ratio = total supply or other purchases/average accounts payable
Average Number of Days Payables Outstanding (non-ratio metric related to accounts receivable turnover ratio)	Average number of days payables outstanding = days in period to track/accounts receivable turnover ratio during period
Inventory Turnover Ratio	COGS/average inventory
Days sales in inventory (non-ratio metric related to inventory turnover ratio)	Days sales in inventory = (average inventory/COGS) X 365
Asset Turnover Ratio	Asset turnover ratio = net sales/average total assets
Fixed Asset Turnover Ratio	Fixed asset turnover ratio = net sales/average fixed assets
Total Assets to Sale	Total assets to sales = total assets/sales
Fixed Assets to Total Assets	Fixed assets to total assets = fixed assets/total assets
Working Capital Ratio	Working capital ratio = current assets/current liabilities

Summary from Netsuite[2]

Operating Efficiency Ratio

The operating efficiency ratio measures how a company uses resources to generate revenue. Similar terms are efficiency ratio, operational efficiency ratio, operating efficiency ratio and operating ratio.

Operating efficiency or operating ratio = Expenses (OPEX or operating expenses + COGS (cost of goods sold)/Net Sales

Accounts Receivable Turnover Ratio

It tracks how a company is doing with customer payment collections in accordance with the terms.

Accounts Receivable Turnover ratio = Net Sales/Average Accounts Receivable

Average Collection Period

The average collection period shows the average number of days between a sale on credit and when the customer pays.

Average Collection Period = Days in tracking period/Accounts receivable turnover ratio in tracking period

"You can't do much with these two—Accounts receivable turnover ratio and Average collection period—but wait until we get to the accounts payable ratio," says Maria, in anticipation of the "salty" comments and questions from Jon.

Accounts Payable Turnover Ratio

This ratio shows how fast a company pays its suppliers and other third parties considering a certain period of time.

Accounts payable turnover ratio = Total Supplier and outside party purchases in tracking period/Average accounts payable

Where

> **Average accounts payable = (Beginning balance of accounts payable in tracking period + Ending balance of accounts payable in tracking period) /2**

Average Number of Days Payables Outstanding

"Anyone can help me with the definition?" asks Maria.

"Easy peasy, we've seen average collection period…. It's the same, but with suppliers instead," says Tim.

Tim is right. The average number of days payable outstanding shows the average number of days between a purchase on credit or a bill and when the company pays it.

> **Average number of days payables outstanding = Days in tracking period/ Accounts payable turnover ratio in tracking period**

Inventory Turnover Ratio

"Aha! Here we go again with inventory!" Ruben says.

"With these bits and pieces, the inventory part will be like Febreze for you!" says Maria.

This ratio shows if the inventory is moving and how quickly it is doing so.

> **Inventory Turnover Ratio = Cost of goods sold in tracking period/ Average Inventory**

Where

> **Average inventory = (Beginning Balance of Inventory in tracking period + Ending Balance of Inventory in tracking period) /2**

"How are we gonna do this with the hundreds or thousands of part numbers? It's impossible," says Jon, shaking his head.

"This may work for mom-and-pop companies but not for us. We are large," supports Mike.

"Got you both. I don't want to get ahead too much, but I also want to answer your questions. We categorize the items and prioritize. We don't do this with every single part number," explains Maria, calming down the rough waters.

Days Sales in Inventory

This ratio measures how many days it takes to turn inventory into a sale. It shows how many days to wait—like the countdown on New Year's Eve—for the magic to happen, for the spell to take place.

$$\text{Days sales in inventory} = \text{Average inventory} / \text{COGS (cost of goods sold)} \times 365$$

Asset Turnover Ratio

The asset turnover ratio indicates how a business uses and leverages its assets to generate revenue.

$$\text{Asset turnover ratio} = \text{Net sales} / \text{Average total assets}$$

Fixed Asset Turnover Ratio

"You know this one," Maria addresses the team.

Sarah starts answering with a low tone that increases in intensity like a tornado gaining strength: "This ratio shows how a company generates revenue from its fixed assets. I think the formula is Net sales divided by Average fixed assets."

"Way to go, girl!" Maria says, giving Sarah a high five.

"Fixed assets are machinery, equipment, things like that, right?" asks Tim.

"Yeah, Tim! Another high five," Maria says and writes the formula on the board.

$$\text{Fixed asset turnover ratio} = \text{Net sales} / \text{Average fixed assets}$$

Where Fixed assets = plants, machinery, equipment, etc.

Total Assets to Sales
"This is the other side of the same coin. You need to flip the formulas," clarifies Maria.

$$\text{Total assets to sales} = \text{Total assets}/\text{Sales}$$

Fixed Assets to Total Assets
This ratio shows the company's total assets relative to its plant, equipment, machinery and the like.

$$\text{Fixed assets to total assets} = \text{Fixed assets}/\text{Total assets}$$

Working Capital Ratio
This is a fundamental ratio. Also known as the "current ratio," this ratio measures the company's ability to pay off its current liabilities with its current assets.

$$\text{Working capital ratio} = \text{Current assets}/\text{Current liabilities}$$

Nanoseconds after showing the working capital ratio to the team, Maria sees a Team message popup on her screen. It's Ernesto.

"How are you doing? How's everything going with the team?" reads Maria.

"All is going well, Ernesto. We have covered financial statements, and we are now covering metrics with focus on the efficiency ratios, where supply chain and procurement have a massive impact," Maria replies.

"Good! Just make sure all is actionable. I want to see execution to get results," highlights Ernesto.

"Of course, Ernesto! It's all about action and results," ensures Maria.

"Let me know if you need anything from us. I need to hop on a call. Talk to you later." Ernesto ends the conversation.

To regroup, Maria tosses a Chips for Thought, like throwing a frisbee to get the team's attention back.

Chips for Thought

1. In what kind of financial ratios do supply chain and procurement have the largest impact?
2. Are you aware of the baseline or the current levels of any of these ratios?
3. What key ratios are you impacting with your current projects?
4. Can you quantify the impact of your work on these ratios?
5. If you cannot quantify the impact, what are the challenges, and what data would you need?

With Maria's lead, the team spends a good amount of time answering these questions. The outcome shows a weak link with the finance and accounts payables team. This is a start, and there's more work to do ahead.

Investment Valuation or Market Ratios

"C'mon Maria, this is pure finance," says Jon, scratching his head when he sees the investment valuation and market ratios words.

Maria clarifies that the plan is to do a quick overview of the key ratios—like "cleaning" the house before your in-laws visit—as CPG companies are using M&A[3] (mergers and acquisitions) to add brands to their portfolios and boost manufacturing considering supply chain disruptions.

Once upon a time—not many years ago—CPGs avoided owning manufacturing facilities, and some reduced their manufacturing assets. The preference was focusing on innovation and keeping relevant existing products instead of investing time in workers, machinery, and overhead.

Fast-forward to today, CPGs buy brands together with the associated manufacturing capacity.

"I read the news that Utz brands acquired Festida Foods last year to improve the supply chain of On the Border brand," comments Mike.

"Festida Foods…" says Jon, rubbing his head, "is that the largest manufacturer of tortilla chips?"

"Yes, it is," Maria answers and adds, "This article from Food Dive names quite a few brand and manufacturing acquisitions like Dot's Homestyle Pretzels by Hershey and Planter's from Kraft and Heinz by Hormel."

"Let's get on these market ratios," suggests Maria.

Price to Earnings ratio
It measures the company's current share price compared against its per-share earnings. A large P/E means that investors expect higher earnings in the future. In this situation, the stock is overvalued. Likewise, a low P/E ratio means lower earnings in the future, and the stock is undervalued.

$$P/E \text{ ratio} = \text{Market value per share/earnings per share}$$

Price/earnings to growth ratio (PEG ratio)
This ratio adds the expected growth earnings factor into the calculations. Investors also use this ratio to determine the actual stock's value. A high PEG indicates that the stock is overvalued. A low PEG is the opposite; the company is undervalued.

$$PEG \text{ ratio} = (price/ \text{ earnings per share})/\text{earnings-per-share growth}$$

Price to sales ratio
This ratio measures how much the market values every dollar of the company's sales. In a different way, how much a person is willing to pay to buy one

share in comparison to what that share generates in revenue. A high ratio means overvalued stock while a low ratio translates into undervalued stock.

$$\text{Price to sales ratio} = \text{market value per share}/\text{sales per share}$$

Or

$$\text{Price to Sales Ratio} = \text{Market capitalization}/\text{Sales}$$

Price to book value ratio
This ratio shows the company's stock price or market value relative to the book or net asset value.

$$\text{P/B Ratio} = \text{Market price per share}/\text{Book value per share}$$

Where

$$\text{Book value per share} = (\text{Total common shareholders equity} - \text{preferred stock})/\text{\#Common shares}$$

Or

$$\text{Book value per share} = \text{Net asset value} = (\text{Assets} - \text{liabilities})/\text{Number of shares outstanding}$$

Metrics in CPG for benchmarking

"Maria, you've dumped a lot on us" says Mike.

"I'm lost. I don't know what to do with all this information," says Tim, sharing the same sentiment.

"I know, guys. It takes some time to digest. But I have an activity that I think helps," Maria says while digging into a big brown carton.

"Ay, ay, ay!" says Ruben, "wouldn't a tequila be better, with lots of salt?"

"Or margaritas!" suggests Amy.

"I don't have tequila or margaritas…I have a game on metrics," says Maria, pulling out a deck of cards after her search in the mysterious big brown box. Maria deals the cards while the team members look at her as if they are getting snow on the Fourth of July.

The cards contain questions about metrics, including purpose, formulas, and examples. The game creates a competition that facilitates learning. Maria has also played the same game—with some adjustments—virtually. The team members in both situations—in-person and virtual trainings—advance in the learning process and have fun with it!

The revised Bloom's taxonomy has the following levels—from basic to advanced:
1. Remembering
2. Understanding
3. Applying
4. Analyzing
5. Evaluating
6. Creating

With her trainings, Maria aspires to take the team at Alex's Snacks to the levels of applying (execution) or analyzing—as minimum—per this revised Bloom's taxonomy. Based on the pyramid, the applying level represents about 50% of the full path. But Maria sees a hungry team. They are hungry not only for potato and tortilla chips (who wouldn't be?), but also for high performance—and this is critical to develop an empowered and effective team.

"Ready for more?" asks Maria.

"Bring it on," says Tim.

Common Metrics and Levels in CPG

"We have covered the key financial ratios to identify and quantify our impact. I'd like to go over what levels you can expect based on other CPG companies. How does this sound?" asks Maria.

"Interesting," says Mike.

"Let's do it," Amy concurs.

CPG companies enjoy high gross margins (revenue less COGS). They sell a high volume of products; this means that their top line in the P&L is high. As their volume is significant, they can take advantage of economies of scale and lower their COGS.

In CPGs, accounts receivable and inventory activity ratios are key areas for assessment, as CPGs have retailers and wholesalers—including large grocery stores like Walmart, Target, Meijer, Costco, Sam's Club, etc.—as customers. CPGs also have and develop ecommerce channels.

The S&P 500 includes the following CPGs:
1. Procter & Gamble (PG) household and personal products
2. Coca-Cola (KO) beverages
3. PepsiCo (PEP) non-alcoholic beverages and convenience foods
4. Philip Morris International (PM) tobacco
5. Mondelez International (MDLZ) snacks
6. Altria (MO) tobacco
7. Colgate-Palmolive (CL) household and personal products
8. Kraft Heinz (KHC) packaged foods

"I think that PepsiCo and Mondelez International are good for benchmarking," says Sarah.

"Agree with Sarah, but how are we going to get the data?" asks Jon.

"We'll send you, Jon, to spy," says Ruben, pointing his index finger at Jon.

"No need," says Maria, still laughing about Ruben's joke. "They are public companies, so their financial information is available."

For CPG companies, it is important to look at the following ratios:

Profit Margins
CPG companies have high margins. **Gross margin** is revenue less COGS. Gross margin ratio is the gross margin divided by revenue to get the percentage. By looking at Mondelez International, their gross margin percentage is 37.10% while it is 53.53% for PepsiCo.

Because of economies of scale, CPGs have lower COGS. COGS includes all production costs, regardless of the manufacturing plant's ownership. Alex's Snacks works with co-manufacturers. The costs incurred with co-manufacturers go to COGS too.

When companies have their own manufacturing facilities, the production costs are in COGS. The difference is that PPE shows in the balance sheet under Fixed Assets—this is Capital Operating Expenses or CAPEX—and the depreciation into the P&L.

Operating margin is the gross margin less the operating expenses or OPEX. The operating margin ratio is the operating margin divided by revenue. Examples of OPEX are research and development (R&D), wages and salaries, overhead costs (not factory) such as selling, general, and administrative expenses (SG&A).

"Curious. What's the difference between Mondelez and PepsiCo in operating margin?" asks Amy.

"Good question, Amy," says Mike.

"Mondelez has 15.93% and PepsiCo 14.40% in operating margin," answers Maria by looking at the Market Watch website[4] on her screen.

"It seems that there is a big drop for PepsiCo," observes Sarah, "the margin goes from over 50% to 14-15% adding the operating expenses."

"Do they know what they're doing?" asks Jon, holding his head up high.

"Well, Mondelez beats them. Their gross margin is 37% and their operating margin is 16%, higher than PepsiCo that starts with a 50% gross margin. Unbelievable!" says Tim.

"That doesn't mean that PepsiCo is doing a terrible job compared to Mondelez," says Maria, making eyeballs pop out. "It depends on the business strategy."

Companies may decide to invest in R&D and diversify their portfolio. This means higher operating expenses and lower operating margins for a period of time. Another situation could be the implementation of processes, systems, and training that show as high operating expenses in the short term.

It is important to analyze the values of the ratios and validate alignment with the strategy. This strategy alignment is what matters. "Okay," says Ruben with a grim look, "PepsiCo doesn't suck." After the inevitable pause, Maria continues talking about the net margin ratio.

Net margin is the operating margin less all other expenses—including the business's capital expenses and taxes. This is the bottom line. The net margin ratio is the net margin divided by revenue.

"Tell us about PepsiCo and Mondelez, Maria," says Amy.

"That's what we wanna know," adds Mike.

"Drumroll please," says Maria. "Mondelez has 14.97% of net margin percentage and PepsiCo has 9.59%."

"I think we have a winner!" says Ruben.

Teasing aside, it's fundamental to listen what the data tells. Companies monitor ratios over time and benchmark against their peers. A value or percentage may look ugly. But it may not be bad, when it is aligned with the strategy.

Metrics

In addition to profit margins, CPG companies look at activity or operating or efficiency ratios. For instance, their accounts receivable may be high, as they sell to retailers with several locations or "doors" nationwide. The same applies to inventory. Accounts receivable management and inventory management are two critical areas for CPG companies, and Alex's Snack is not the exception.

Key performance metrics in these two areas are as follows:
1. Accounts receivable days
2. Inventory turnover
3. Inventory days

"It looks like we need to watch out for inventory no matter what," says Tim. "Yes, Tim, as a dog watches and protects a raw beef bone," replies Maria.

Before getting into the strategic goal tree and SCOR model, Maria tosses another Chips for Thought.

Chips for Thought

1. From the summary table with the activity ratios above, which ones would you select to track? Why?
2. What are the levels of those ratios in the companies you are benchmarking against? Hint: you may check Market Watch website
3. How do the values determined in question 2 of this Chips for Thought compare with your company's values?

Strategic Goal Tree and SCOR model
Back in 2003, during her time at Michigan State University (MSU), Maria saw a diagram showing the impact on ROI that struck her. Boom! That was

the visual to understand how powerful supply chain and procurement are on the financials of a business. Since that moment, Maria and the diagram have become best friends forever.

Source: Michigan State University

The diagram displays the ROI calculation in a non-conventional way, linking the P&L or Income Statement with the Balance Sheet. Simple and powerful. Maria asks the team at Alex's Snacks to look at the diagram, as much a masterpiece for supply chain and procurement as the *Mona Lisa* in the art world.

"How do we include supply chain and procurement metrics aligned to those in finance and to the business?" asks Maria. "Here comes the concept of the strategic goal tree." The first time that Maria heard about this concept was while she was working at Cummins Inc. headquarters in Columbus, Indiana (a lovely town much smaller than Columbus, Ohio).

It is a strategic goal tree because it has wide branches—net income and total assets in the ROI diagram—that extends to smaller branches. For instance, the net income branch has the smaller branches of sales and total expenses; the total asset branch has current and fixed assets as smaller branches.

These smaller branches have, in turn, their own branches. Following the example in the diagram, sales has price and quantity while total expenses has COGS and other expenses. COGS has three branches: raw materials, labor, and fixed costs. "From here we can add supply chain and procurement metrics, growing the tree with new branches and the ROI too," says Maria.

When companies scale up, there are more team members and new teams. The goal tree becomes bigger, and like a big tree, it sometimes needs a trim. Pruning a tree helps to keep it healthy and growing at its best. A strategic goal tree with too many branches or metrics creates confusion, and it is more challenging to manage.

There are various reasons to prune a tree. It's a good practice to make it a habit, an article on getbusygardening.com[5] indicates. But the same article states that most people think about trimming trees after the damage of a storm, when the lower branches are getting in the way or there is potential for harm. Other reasons include fostering flowers and fruit production, to trigger new growth.

The same applies when reviewing the metrics (branches) of the strategic goal tree. Companies may assess it after a major event only or when it kind of gets out of control situations like supply chain disruptions, brand or company acquisition, or fast growth. As getbusygardening.com suggests for the best practices with trees, a proactive periodic review is better.

Companies understand the importance of pruning the strategic goal tree at regular intervals for growing their best. Like with any other tree, trimming the strategic goal tree requires a series of steps.

Steps to follow:
1. **Any tree:** Trim off any suckers growing at the base of the trunk.
   Strategic Goal Tree: Remove any metrics that are not in alignment with the strategy for a greater ROI.
2. **Any tree:** Remove all the dead or dying branches.

Strategic Goal Tree: Remove all metrics that the team doesn't use and which create noise only.
3. **Any tree:** Prune out unwanted or hazardous branches.
   Strategic Goal Tree: Remove any metrics that do more harm than good.
4. **Any tree:** Remove any damaged or weak branches.
   Strategic Goal Tree: Make any metric earn its keep; otherwise, sweep it away.
5. **Any tree:** Trim out overlapping branches that rub together.
   Strategic Goal Tree: Remove any metrics that overlap. Keep the tree healthy for profitable growth.

SCOR DS Model

"Let's add the procurement and supply chain branches to our strategic goal tree," suggests Maria.

"Do we need to go through this exercise?" Jon asks. "I'm sure the model looks good on paper, but I don't know if it helps here. I don't think this is what Ernesto wants either."

"Understood, Jon. The objective with this is to have supply chain metrics aligned with the financial metrics and see how we are doing against peers and competitors," says Maria. "Who knows? We may be outperforming Nestle, PepsiCo and Mondelez."

"I'm sure we do," says Ruben with a voice sounding like the witch planning Snow White's death with a poisoned apple.

"Let's see," says Maria, piquing their curiosity and interest.

The SCOR DS stands for Supply Chain Operations Reference Digital Standard Framework. Most people refer to it as SCOR. The SCOR model started in 1996. At that time, the management consulting firm PRTM and AMR (now part of Gartner) built the SCOR model. The Supply Chain Council, an organization acquired by APICS in 2014, promoted the model. In 2019 APICS changed its name to the Association for Supply Chain Management (ASCM).

ASCM allows free access to the SCOR model on their website: https://scor.ascm.org/performance/introduction[6]

In late 2022, a group of diverse experts made a major revamp of the model with updates to processes, metrics, skills, and practices. The updated metrics provide businesses with new ways to measure and improve their supply chains.

The WHY of SCOR

"Love the history, Maria, but what is it?" asks Jon with an impatient tone.

"The why first, Jon, and then we move into the what, would that be okay?" asks Maria.

Paraphrasing what SCOR indicates online, Maria explains that the goal is to develop supply chains that focus on satisfying customer demand.

Companies like Alex's Snacks can use SCOR to analyze their strategy and quantify their performance, as the SCOR framework is robust in performance metrics and practices. Businesses can apply SCOR to benchmark, evaluate their supply chain performance and identify targets for improvement. SCOR mentions four main techniques:

1. Continuous supply chain improvement, by mapping, measuring, assessing, and validating processes and activities.
2. Supply chain performance comparison at every level and benchmark, like-for-like measure to identify gaps and determine actions to take for performance improvement.
3. Supply chain current practices' evaluation to identify practices and technology that lead to higher performance levels.
4. Organization's design with focus on continuous learning and skill development centered in performance.

"Any questions about the why?" Maria asks. "Going once, going twice…let's go over the what and how."

The WHAT and HOW of SCOR

"I wanna see how we're doing against the big guys," says Mike.

"We're getting there, Mike," says Maria, "sit tight and buckle up because the SCOR model is a ride with loops. Literally, loops." She then shows the SCOR graphics, a double-infinity diagram, that shows "the looped, continuous, and connected nature of today's supply chain," per ASCM.

This is the link—https://scor.ascm.org/processes/introduction[7].

In its framework, SCOR considers 7 core processes in the supply chain:
1. Orchestrate
2. Plan
3. Order
4. Source
5. Transform
6. Fulfill
7. Return

"Can you see these processes in the diagram?" asks Maria, gazing at the team members' faces. "The horizontal infinity loop shows the demand-supply balance and the vertical infinity loop links synchronize and regenerate."

Not to bore Jon with history, but Maria explains that all previous versions the SCOR model had the usual linear representation. The 2022 update shows the supply chain as interconnected networks.

This version also adds "orchestrate" to show the importance of strategy, business rules, technology and human resources moving in alignment, like musicians following the direction of the conductor with her baton and bodily movements. Other changes of the model are the split of "Deliver" into "Order" (customer orders) and "Fulfill" for better focus on these activities, practices, and metrics. "Make" becomes "Transform" to cover more types of manufacturing and service providers.

Structure of the SCOR framework
There are four building blocks:
Performance
"If you are multi-tasking, come back to me," says Maria, "here are the supply chain and procurement metrics we're looking for." This section contains the standard metrics to measure process performance and to define strategic goals—the branches to add to the strategic goal tree.

Processes
It contains the standard descriptions of management processes and process relationships.

Practices
Practices include best practices that lead to higher levels of performance.

People
The people section indicates the standard definitions for skills required in the different supply chain processes, including suggestions for training.

"For the visual learners in the room, you can see the four building blocks on the ASCM website[8]," says Maria.

She adds "SCOR is not a bunch of lists (or endless lists) like your to-do list or the long list of emails you need to read; it's a framework that combines these four building blocks and adds pre-defined relationships between material, information, and workflow processes." For example, performance metrics are linked to specific processes and practices. ASCM estimates that SCOR comprehends 95% of most business flows in supply chain.

Our Beloved Performance Metrics
The performance metrics is the most important section. Although SCOR DS is a process framework, ASCM recommends companies start with the metrics because they are like lightbulbs blinking when something is wrong. Metrics assists in identifying the gaps or fires to extinguish. "It's like kind of you're ready to fix, but you need to know what," summarizes Ruben.

The performance section of SCOR includes performance attributes, metrics, and process or practice maturity. These factors are different from the levels in the process and metrics hierarchies.

"Is everyone on the SCOR DS site?" asks Maria. "When you click on performance in the top menu, you see introduction, resilience, economic, and sustainability. Are you with me?"

"Got it," says Tim.

"Me too," says Amy.

"Ruben, Sarah, Jon, okay?" Maria asks to ensure each member of the team is on board and in the same boat.

Resilience, Economic, and Sustainability are three areas of performance. There are performance attributes—a category of metrics used to express a specific strategy—in each of the three performance areas. They are as follows:

Resilience
1. reliability
2. responsiveness
3. agility

Economic
1. cost
2. profit
3. assets

Sustainability
1. environmental
2. social

"Hit the pause button, Maria," says Jon, like in the movie Click where Adam Sandler has a TV-like remote control to make changes in reality. "Can we see this in real life, not from a top-notch research website?"

"Of course," Maria replies, and follows with an example. Per the conversation with Ernesto, Alex's Snacks is growing so fast that meeting demand is a challenge. Because of this, there are lost sales. In the SCOR DS model, this situation relates to resilience and within resilience, reliability is the best fit.

Another example is the lower profit margin challenge that the company faces. In this case, the performance area in the framework is economic and corresponds to profit and cost performance attributes.

"And so what?" asks Jon with a face like he is drinking coffee with lots of potato-chip salt in it.

"So far we have the financial metrics and the supply chain performance attributes or categories based on the strategy. We now need our beloved supply chain and procurement metrics," explains Maria, with a feeling of butterflies in her stomach, as if she sees the stars aligned in the sky.

Metrics measure the ability of processes to achieve the strategic objectives of the 8 corresponding performance attributes:
1. Reliability
2. Responsiveness
3. Agility
4. Cost
5. Profit
6. Assets
7. Environmental
8. Social.

In the SCOR metric hierarchy, there are three levels of pre-defined metrics:
1. Level 1 metrics—This level is for strategic metrics like key performance indicators. They help to set targets aligned with the strategy.
2. Level 2 metrics—Level 2 is the child metrics of Level 1; or level 1 is the parent metrics of level 2. Level 2 metrics work as the diagnosis of level 1 metrics in identifying root causes of a performance gap of level 1 metrics.
3. Level 3 metrics—Level 3 metrics operate in the same way as level 2 in reference to level 1 metrics. Level 3 works as a diagnosis of level 2 metrics.

"Isn't this beautiful?" asks Maria, displaying a big smile. "Having it all mapped out, with the financial and supply chain metrics aligned based on strategy?" Maria feels like a chorus from Heaven is singing "Hallelujah."

"It's gorgeous," says Amy.

"I like that the framework takes a quantitative and analytical approach, not like others that are…"

Sarah stops speaking as Ruben interjects, "Say it, Sarah, you want to say that they are BS." Sarah blushes.

"I don't remember seeing this model at school. It looks solid," Tim says.

"Can we compare against the competition? How are those guys doing?" Mike asks.

"You're ahead, but that's where we are heading now," says Maria.

Maria reassumes the challenges of lost sales and lower profit margins to apply the SCOR model. With lost sales, we look into the resilience area and the reliability performance attribute.

The level 1 metrics that measure whether processes accomplish reliability are as follows:
- RL1.1 Perfect customer order fulfillment
- RL1.2 Perfect supplier order fulfillment
- RL1.3 Perfect return order fulfillment

Taking the level 1 metric listed in the first place—perfect customer order fulfillment—Alex's Snacks needs to track the percentage of orders meeting delivery performance expectations, including complete and accurate documentation and no delivery damage (perfect quality).

The calculation of the perfect customer order fulfillment = (Total perfect orders/Total number of orders) x 100%.

"What happens if the order has multiple items? Everything needs to be perfect?" asks Amy. Per ASCM, "An order is perfect if the individual line items making up that order are all perfect."

"Another question," starts Tim, "I know how to get the number of orders but how do we get the total of perfect orders?"

"Excellent question! We need to go down the hierarchy to get to the level 2 metrics," answers Maria.

The level 2 metrics under perfect customer order fulfillment are:
- R.L.2.1 – Percentage of orders delivered in full to the customer
- R.L.2.2 – Delivery performance to original customer commit date
- R.L.2.3 – Customer order documentation accuracy
- R.L.2.4 – Customer order perfect condition

These are the metrics that work as the diagnosis of the perfect customer order fulfillment and identify the performance gaps. By tracking these level 2 metrics, Alex's Snacks can pinpoint whether the issues are related to completing the order quantities (R.L.2.1), or meeting the delivery dates (R.L.2.2), or having the documentation correct (R.L.2.3), or satisfying the quality standards (RL.2.4).

"And with these measures we do the calculations for the level 1 metric. Is that right?" asks Tim.

"You get it, Tim!" answers Maria, and explains that the SCOR model has level 3 metrics to go deeper. They work as a diagnosis of level 2 metrics like the percentage of orders delivered in full to the customer (R.L.2.1) or delivery performance to the original customer commit date (R.L.2.2).

SCOR contains the definitions of the performance attributes and thorough explanations of the set of metrics. It uses a code containing letters and numbers for each performance metric that starts with the associated performance attribute—for example, RL, RS, AG, CO, PR, AM, EV, and SC, followed by a numeric series to show the hierarchical level of the metric.

"Are we there yet?" asks Mike, laughing and waiting for an answer from Maria.

With no need for further explanations, "Yes, we are," she says, and continues to explain that the performance section of the SCOR model helps Alex's Snacks to understand its performance relative to competitors. ASCM has SCORmark Benchmarking[9] that combines the SCOR Digital Standard hierarchy with PwC's (PricewaterhouseCoopers) historical data from more than 1,500 organizations and 2,500 supply chains to help identify gaps and make improvements.

"So, Mike," says Maria, "the competitive benchmarking and the expected performance of process KPI's compared against the actual performance of you guys, Alex's Snacks, shows the gaps in performance and works as a roadmap for continuous improvement."

"Can't believe that I'm saying this…can we do a Chips for Thought?" Mike asks. "I'd like to know where we stand."

"That's a Yes or Yes question," says Maria.

Chips for Thought

1. Based on your company's strategy, select an area of the performance section in the SCOR model and map out level 1, level 2, and level 3 metrics.
2. Map out the supply chain metrics with the financial metrics.
3. Calculate the current levels of each metric.
4. Benchmark against competitors to identify the gaps.
5. Show the plan to cover these gaps and for continuous improvement.

Processes in SCOR

SCOR also has a process section. The framework assists companies with quantifying the gaps in service, cost, and inventory between the as-is and the to-be state of the business and the supply chain. As there can be many—maybe too many—potential improvement processes, companies prioritize the processes, practices, and performance metrics to achieve the to-be state.

The continuous improvement journey goes from the as-is to the what-if to arrive at the to-be state. SCOR provides tools and resources to assist along the way as follows:

1. Document what the business and supply chain are doing and their organization. This is the what of the as-is.
2. Document the capabilities of the business and enterprise supply chain. This is the how of the as-is.
3. Document and test alternatives to organize the business and the enterprise supply chain. These are what-if scenarios.
4. Document the desired organization of the business and supply chain. This is the what and where of the to-be.
5. Document future processes. This is the how to get to the to-be.

Like with the performance metrics, there is a hierarchy of the processes. SCOR DS identifies seven major processes. "Team, what are these seven key processes?" asks Maria.

"Here," says Tim in a rush. "They are orchestrate, plan, order, source, transform, fulfill, and return."

"Epic! Like my kids say," Maria indicates.

Orchestrate is level 0 in the model as it is in a strategic level. The other 6 major processes are at level 1. Level 2 is process categories. They show the capabilities that enable level 1 processes. Level 3 is process elements. They are a sequence of steps to plan, source, make, deliver, and return. Level 4 is process improvement activities. As there are several, standardization becomes challenging. Because of this, SCOR doesn't make recommendations.

Practices in SCOR

Per ASCM, a practice is a unique way to configure a process or a set of processes. It can be unique because of an automation, or a defined sequence, or a technology or skill applied.

SCOR practices help companies with the following:
1. Standardizing processes
2. Identifying alternative methods of operating the supply chain
3. Formulating a wish list of process configurations and opportunities for automation
4. Formulating a "blacklist" of undesired processes configurations

SCOR practices span level 1 and level 2 SCOR processes around two practice pillars:
1. Level 1—These practices extend to all the supply chain. Best practices are the only level 1 in SCOR.
2. Level 2—These practices cover performance of a specific level 1 practice or best practices. They are about 280.

With so many practices, selecting which practices are best for a company's particular situation is challenging. To address this, SCOR provides a simple methodology using a quadrant grid and push pin for each SCOR practice. Based on the return and amount of effort or risk, the quadrant includes quick wins, nice to have, sponsor issue, and consider carefully.

"How about another Chips for Thought?" asks Maria.

Chips for Thought

1. Consider the list below with all the practice categories and select the most helpful practice categories for your organization.

Categories
✓ Business Process Analysis and Improvement
✓ Customer Support
✓ Distribution Management
✓ Information and Data Management
✓ Inventory Management
✓ Manufacturing and Production
✓ Material Handling
✓ New Product Introduction
✓ Order Engineering
✓ Order Management
✓ People Management (including training)
✓ Planning and Forecasting
✓ Product Lifecycle Management
✓ Purchasing and Procurement
✓ Reverse Logistics
✓ Risk and Security Management
✓ Sustainable Supply Chain Management
✓ Transportation Management
✓ Warehousing

2. Include the practice categories selected in the quadrant and decide which are the top three to pursue.

SCOR DS Model Limitations

"The SCOR model is GREAT! Amazing!" says Maria, and Ruben says, "Epic!" Like the song "Can't Take My Eyes Off of You" says, this is too good to be true. The SCOR model has its limitations:
1. SCOR doesn't address sales and marketing, including demand generation, product development, and research and development.
2. SCOR doesn't include which process improvement activities companies need to implement. This is level 4 in the process section. Some process

improvement techniques include kaizen, lean, TQM, six sigma, technology, etc.

"If I get this right, we map out all the metrics, identify gaps or processes underperforming, know the best practices and then we don't know what to execute to improve because this is a limitation of the model," says Jon. "What's the point of all this? To look good on paper, as I said from the beginning." Jon complains and answers his own question.

"At Alex's Snacks, we are going to execute. Period," clarifies Maria. "We need another framework that complements the SCOR model for continuous improvement. Have you heard about lean six sigma?"

Endnotes

1. Rosemary Carlson, *What are Solvency Ratios?*, The Balance, 2020, https://www.thebalancemoney.com/what-are-solvency-ratios-and-what-do-they-measure-393211.
2. David Luther, *Top Efficiency Ratios: Operational, Asset, Inventory and More*, NetSuite, 2022, https://www.netsuite.com/portal/resource/articles/accounting/efficiency-ratios.shtml.
3. Christopher Doering, *How CPGs are using M&A to boost manufacturing amid supply disruptions*, Food Dive, 2022, https://www.fooddive.com/news/cpgs-mergers-acquisitions-manufacturing-supply-disruptions/627838/.
4. MarketWatch, https://www.marketwatch.com/investing/stock/mo/company-profile.
5. Amy Andrychowicz, *How to Trim Tree Branches Yourself: A Step-By-Step Pruning Guide*, Get Busy Gardening, 2022, https://getbusygardening.com/how-to-trim-tree-branches/.
6. ASCM, *Introduction to Performance*, https://scor.ascm.org/performance/introduction.
7. ASCM, *Introduction to Processes*, https://scor.ascm.org/processes/introduction.
8. ASCM, *Introduction to Processes*.
9. ASCM, *SCORmark Benchmarking*, https://www.ascm.org/membership-community/corporate-membership/scormark-benchmarking/.

CHAPTER 4:

Process Mapping, Matrices, Frameworks, and Toolkit from Lean Six Sigma

Key Takeaways
In this section, you will learn:

1. The why of lean six sigma and fundamentals
2. Understand and apply the different tools along DMAIC and DMADV including process mapping—why, when, and how
3. Define a solid measurement system (this is critical for alignment)
4. Build Cause and Effects Matrix (C&E); Failure Mode and Effects Analysis (FMEA) matrices and other tools in the lean six sigma toolkit to use in cross-functional processes
5. Build a toolkit to apply lean six sigma to procurement and supply chain for continuous improvement

The Why of Lean Six Sigma and Fundamentals
Lean six sigma is a powerful combination of management methods based on the six sigma principles with a focus on efficiency. Both lean and six sigma pursue more efficient processes that result in higher profit margins. The difference between these two guys is the how.

Six Sigma

Six sigma aims to reduce variation and defect rates in production processes through statistical analysis. Within six sigma, there are two approaches—each has 5 steps:
1. DMAIC (define, measure, analyze, improve, control) and
2. DMADV (define, measure, analyze, design, verify).

"DMAIC is for challenges for which you *don't* have a solution," explains Maria. "If you know what to do, as the Nike slogan says, Just Do It." Maria continues to explain that all start with the current processes or the as-is in the SCOR model, to then identify and execute a solution that addresses the problem.

It is important that the team maintains the solution over time. "It is not efficient to improve a process, get the desired results, and in a few months to come back again to the starting point," says Maria. She also adds that to avoid losing the improvements made, companies perform audits to ensure that all the great solutions are still in place.

The 5-step DMAIC approach works like magic for supply chain performance issues or to adjust existing processes. With entire new functions or processes, the DMADV is the way to go.

The 5-step DMADV approach allows for the creation of new tools. Both approaches, DMAIC and DMADV, share the same first stages and take different paths after the analyze phase. DMADV follows with design and verify. Design is for the development of a whole new solution or tool. Verify is for ensuring that the solution works. Six sigma with DMAIC and DMADV consists of monitoring the supply chain for defects, identifying issues, and solving them in a brutally effective way.

Lean Method

The objective of the lean method is to eliminate waste, providing maximum value to customers with the lowest possible amount of investment, as Purdue University indicates.[1] It started at Toyota, as a business philosophy to achieve the highest levels of efficiency. These are enterprise-wide efforts. Compared against six sigma, the scope is broader than manufacturing.

Process Mapping, Matrices, Frameworks, and Toolkit from Lean Six Sigma

Lean six sigma gets the best of six sigma and lean to have an impactful toolkit for waste reduction. The DMAIC method from six sigma is a structured approach that fosters cross-functional collaboration to eliminate the eight types of waste that lean highlights:

1. Defects—Products or raw materials that fail to meet the standards.
2. Overproduction—Production is more than demand or more than customer orders.
3. Waiting—Process bottlenecks and downtime.
4. Non-utilized talents—"I can bet on this one. It is when we have great people, but we give them stupid things to do," says Mike.
   "Blunt," says Ruben.
   "When companies don't use or allocate resources in an efficient way," explains Maria in a more polished manner.
5. Transportation—Inefficient shipping methods or going further, the need for transportation at all. This waste relates to waste #3, waiting.
6. Inventory—"What's that?" jokes Ruben. "Never heard of it."
   "Excess or surplus of product or raw materials or packaging," says Maria.
7. Motion—Unnecessary movement of product, people, or equipment.
8. Extra processing—Work in excess; doing more work to provide the customer with the same value.

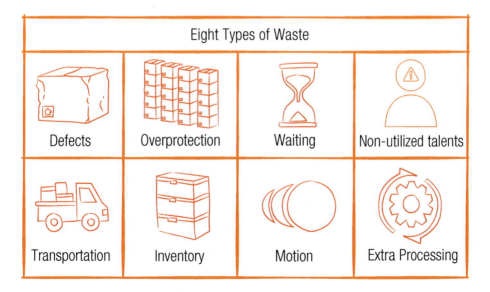

"Any questions, team?" asks Maria.

"Yes," says Sarah, "what's the difference between transportation, motion, and extra processing?"

"You've hit the nail on the head, Sarah. Let's hammer on it," says Maria.

The motion waste occurs within the workstation. It comprehends all movements beyond the bare minimum for completing a process step. It can also be a complete process step.

"Think of looking at multiple systems or folders to get the information you need, like checking for a particular SKU availability," says Maria. "Another example is copying and pasting the same information to build several reports that have multiple formats."

The extra processing or over processing waste refers to any additional features or enhancements that the customer doesn't value. "For example, creating a 300-page report that the customer doesn't read," says Maria.

The transportation waste occurs between process steps and workstations. Like motion and extra processing, the transportation waste is visible or evident. "Forklifts or trucks with pallets of products going from one area to another one, customer returns that are back in the warehouse, are examples of transportation waste," says Maria.

"Ready for a Chips for Thought on lean six sigma?" asks Maria.

Chips for Thought

1. Think of types of waste in the work you do and categorize them per the eight types of waste.
2. Are there any that you can eliminate fast, the low hanging fruit?
3. Are there any that are like weeds—they keep growing after taking action? What didn't work?
4. Listen to a couple of episodes of the "Adventures in Supply Chain" podcast to take note of the key points. Hint: Lean six sigma and circular economy are related. Be prepared to connect the dots!
   Lean Six Sigma—Six Sigma Lean and Circular Economy with Guest Sneha Kumari.[2]
   Circular Economy—Coronavirus and Supply Chain—Circular Economy and Supply Chain with Deborah Dull.[3]
5. What waste would you like to eliminate but don't know how?

Lean Six Sigma Toolkit

"We address a limitation of the SCOR model with the action-packed toolkit of lean six sigma for the continuous improvement of supply chain and procurement," begins Maria, and adds, "This is going to be transformative (and fun!)." Top performance companies follow this approach that extends beyond manufacturing.

"Is this still a valid methodology considering Covid and all the supply chain disruptions?" asks Tim.

"Glad you're raising this concern," states Maria. "If you haven't listened to the episode of the "Adventures in Supply Chain" podcast—also in video format—with Dana Drake-Cox, VP Global Supply of Stanley Black & Decker,[4] I'd recommend you give it a listen or watch it, as Dana addresses this question. Dana indicates that the six sigma approach is valid during these unique times, and she provides fantastic actionable insights." Maria says.

The 5-step DMAIC Approach (Define, Measure, Analyze, Improve, Control)

The DMAIC approach is for existing processes and focuses on the following:
1. Improving performance by flawless execution
2. Achieving rapid breakthrough improvement
3. Applying advanced breakthrough tools that work
4. Making a positive and deep cultural change

"This sounds like Ernesto talking," jokes Ruben. On the big screen, Maria displays the six sigma project chart to provide an overview of the steps and have a visual where the different tools fit.

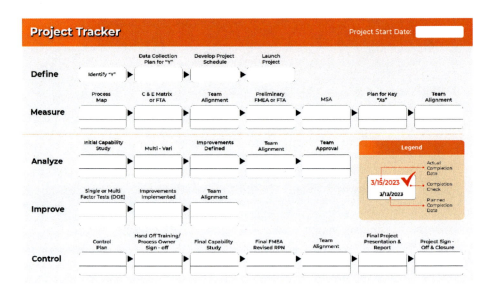

Source: Cummins Inc.

It all starts with a practical problem that converts into a statistical problem in the define phase. This statistical problem becomes a statistical solution facilitated by the toolset. The statistical solution then turns into a practical or executable solution.

Define

"Here is where we translate a practical problem into statistical or measurable terms."

"I wish I could make more money," says Maria, "and Jon gives me a dollar, but is that what I meant? Did I resolve my problem? The same applies when we say we want to lose weight. If we get rid of a pound, did we make it?"

"I don't think so," answers Amy, "we need targets, measures."

"I like where you are heading, Amy," encourages Maria.

Projects in six sigma have a simple equation:

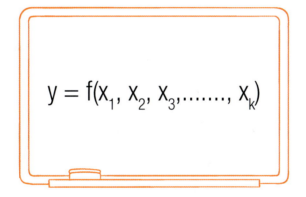

$$y = f(x_1, x_2, x_3, \ldots, x_k)$$

y = the problem
x's = independent variables or factors that impact the y

"Let's start with some examples of the y," suggests Maria.

Example 1:
People say: We spend too much money on expedited freight.
Six sigma says: Reduce the number of expedited shipments.

Example 2:
People say: We have excess inventory.
Six sigma says: Reduce the number of days or weeks or months in inventory.

Example 3:
People say: We are not meeting the customers' required delivery date.
Six sigma says: Increase the number of on-time deliveries.

"We are happy to give you many more problem examples, if you don't have enough with Ernesto's list," says Ruben.

"Great point, Ruben!" replies Maria.

"What?" asks Jon, "what's positive about being inundated with problems?"

"Other CPG companies share the same challenges as you guys at Alex's Snacks," says Maria. "Let me show you some tools that help to select which problems to address."

"Like how to prioritize projects?" asks Amy.

"Right, Amy!" says Maria.

Six Sigma Project Selection Process

"From the SCOR DS model, we have identified our gaps against our historical performance as well as against competitors in CPG. To decide which of these gaps to prioritize, feel free to use this grid," says Maria while she's drawing the vertical and horizontal axis on a whiteboard.

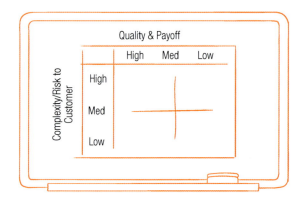

On the y axis, the grid shows the degree of complexity or the risk to the customer. On the x axis, the grid indicates the quality and payoff. It is about the complexity vs. payoff balance. Based on where each project idea is on the grid, the ideas may advance to a six sigma project pipeline.

Companies also use Pareto to select projects. The Pareto principle indicates that 80% of the results or outcome come from 20% of the opportunities or input. This rule helps to identify which project ideas to prioritize.

"I feel it. Another Chips for Thought is coming," says Mike.

"You're right!" says Maria.

Chips for Thought

1. Based on your work with the SCOR model, list the gaps identified or brainstorm potential projects. These are unresolved challenges.
2. Categorize each potential project in the payoff and complexity quadrant.
3. Apply the Pareto principle to your potential projects.
4. Based on #2 and #3, select which projects move on to the six sigma project pipeline.
5. For each project in the pipeline, can you answer these questions? (Source: Six Sigma training at Cummins.)

Questions
1. What criteria did you use in selecting your project?
2. Who is the customer?
3. What data have you collected to understand the customer's requirements?
4. What are the reasons for completing this project?
5. What's the problem you are addressing?

6. Where, when, and to what extent does this problem occur?
7. What are the boundaries of this project?
8. What resources do you need?
9. What do you need from your team?
10. What is the goal of this project?
11. What are the benefits?

Six Sigma Project Description

"You have your project, your baby," says Maria. "Let's create your project description and quantify the expected benefits. Remember, numbers are the language of the business."

A solid project description needs to be clear, like the mountain waters in Canada, Iceland, Upstate New York, or down south in Patagonia. This means that the description includes a measurable y of the project and the following elements—in Amy's words, "We need targets, measures."

1. Baseline—How good is the starting point? 85% on-time deliveries? Measure it!
2. Entitlement—What's your dream? 100% on-time deliveries?
3. Closing the gap—How could you improve it? How can you get higher than 85% on-time deliveries?
4. Goal—How much of the gap do you want to close during this project? Shoot for 95% or 90% or a different percentage?

Writing is thinking. Maria asks the team members to complete the template below with their project description and expected benefits. This request comes with some warnings:

1. Boiling the ocean—"Hit Me With Your Best Shot," as the song by American rock singer Pat Benatar says. It's good that teams make their best efforts to reach gigantic accomplishments. But teams' first shot doesn't need to be at the entitlement.
   There can be more than one best shot. Each time, it's important to set boundaries and define the scope.
   "With the Chips for Thought and all of this about lean six sigma, my best shot is a shot of tequila," Ruben says, making all the team members laugh.

2. Micro-focused—The pitfall here is that the scope is too narrow, or the goals are too small. This is the opposite of boiling the ocean. To avoid this situation, "Shoot for the moon. Even if you miss, you'll land among the stars"[5] as Norman Vincent Peale would put it.

    "As we are talking about shots and shooting, Ernesto and other executives or directors call for the shots on the project scope," Maria says. "That's a good one!" celebrates Tim.
3. Solution in mind—Six sigma doesn't apply when the team already knows what to do. If they do know, what applies is the *Just Do It*, Nike's slogan.
4. Too many metrics—"If you have more than three priorities, you don't have any," says Jim Collins, author of *Good to Great: Why Some Companies Make the Leap and Others Don't*. With a high number of metrics, teams may end up measuring nothing.

"Team, ready to complete the project description template?" asks Maria.

Project Description Template (Project Name)

Objective: Reduce/ optimize/ increase ....(project y)

from .... (current level)

to .... (good level)

for ....(specific area)

while reducing/increasing/holding constant .... (constraining y's)

Benefits:
- Include business results with a description of how to calculate the business impact. The business impact must be in a dollar amount, the way Scrooge McDuck—Donald Duck's uncle—likes it.

- Measure the impact on the customer. There is no room for wishy-washy benefits like customer delight or next-level supply chain like we are pressing a button in an elevator to go up.

Note: business results and impact on the customer must be measurable and tangible.

"Guys, by popular demand…another Chips for Thought," Maria says.

Chips for Thought

1. Write your project description, including business results and benefits to the customer.
2. Compare and validate your project description with other team members.
3. "With all the team members?" asks Jon, "because there are four and it's gonna take time that I need to be actually working."
   "With another team member is fine," clarifies Maria.
4. After review and validation, read it aloud.

Funneling Effect

Before getting into Measure—the phase that follows Define—Maria explains the funneling effect. The funneling effect lets the team know what to expect from this 5-step process.

"Quick recap of what we have," says Maria.

"We have the project description," answers Sarah.

"And we also have the targets and measures," adds Amy.

"Perfect, ladies!" says Maria, "We have our y problem defined in statistical terms and the business results and the benefits to customers quantified. Wonderful!"

"Remember this equation?" Maria asks. "The x's are the variables that impact our y."

$$y = f(x_1, x_2, x_3, \ldots, x_k)$$

Maria goes on to explain that six sigma projects start with several variables or inputs, over 30 at the measure phase. The number then goes down to about 10 – 15 after this phase to get through the analyze phase and arrive at 8 – 10 variables. From there, 4 – 8 variables in the improve phase and 1 – 3 in the control phase.

The funneling effect occurs when the team does this variable elimination from phase to phase based on the impact of each variable on the y. Any solution to the problem or y involves the variables or inputs with the highest impact.

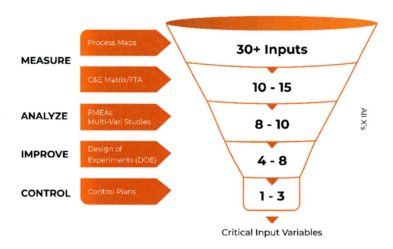

"How does the funnel effect or variable elimination work? What do you use to identify the x's with the highest impact on the y? And what do you do with the variables eliminated?" Amy asks these three questions at once.

Maria puts the project tracker image up again to answer Amy's questions and explain further the phases of measure, analyze, improve, and control.

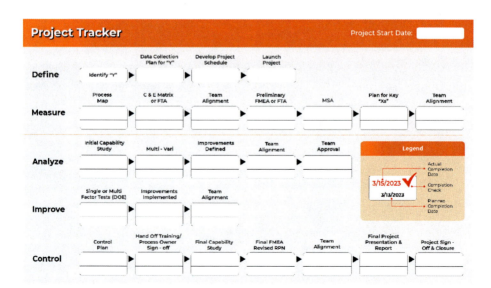

Source: Cummins Inc.

"We can use different tools in each stage that help us prioritize variables. When we make changes, we want them in the inputs with the highest impact to solve our problem, our y," states Maria.

"I see…there are different tools," acknowledges Amy, "and what do we do with the variables that we get rid of?"

"Sorry that I missed that question, Amy," Maria apologizes. "The variables that we discard by the funneling effect we no longer consider, as Taylor Swift sings, we are never ever getting back together. We. Leave. Them. And that's okay, so we can focus on the variables with the highest impact," explains Maria.

Measure

The measure phase starts showing the funneling effect in full force. Like in the Star Wars movies, May the Force Be With You in using these powerful tools: process map, C&E (cause and effects) matrix or FTA (fault tree analysis), the preliminary FMEA (failure mode effects analysis) or FTA, and MSA (measurement system analysis).

"Whew! We're gonna have fun with all these acronyms!" says Mike.

All projects start with a process map that illustrates what happens in the process. It's a helpful tool to define the scope and identify all the potential inputs that could influence the key outputs of the process. If the y is to increase on-time deliveries, the x's are all factors that may impact on-time deliveries.

Process Mapping

Process maps are excellent visual tools that could generate surprises. Like using the Google Maps or Waze apps, it's important to know the current location, the "You're here," on our baseline. In discovering this baseline, "You would be amazed at how many times I've heard I didn't know that we did that or that step in the process existed. Be prepared for the unexpected," Maria warns the team.

There are many types of process maps including the following:

- **Process variables map:** Companies use this type of process map as the first step in the six sigma process (funnel effect). They are critical for process improvement by reducing and removing variation.

Process Map

	Name of Process/Project	
Inputs (Xs)	The "y" of the project	Outputs (Ys)
	C/U/S	
	C/U/S	
	C/U/S	

Buttons: Transfer to C&E | Delete Input (x) | BACK

- **As is/Can be map:** This kind of map is gorgeous for identifying non-value-added steps. "You can note that this is a critical first step in cycle time or lead time reduction projects, like with new product launches or introductions."
- **Swim-lane:** Companies use a swim-lane map when it's relevant to show the teams, groups, departments or organizations that are performing the process steps. A swim lane has three elements: time, job functions,

and tasks or processes. The job functions can be in a column to the left or in a row at the top to then include the different process steps or tasks "in lanes."

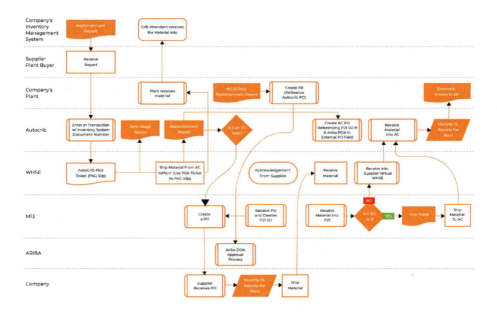

- **SIPOC**: this acronym stands for supplier, input, process, output, and customer. Businesses use SIPOC when there are challenges in scoping. The "triple role" is the concept behind SIPOC.
  The triple role means that each team holds three roles:
  1. The customer receives inputs like physical products and information.
  2. The processor converts the inputs into deliverables or outputs.
  3. The supplier provides these outputs to customers.
  "For example," Maria says, "you, as the procurement and supply chain team, are a customer of product development or engineering that provides information on the specs for film and the timeline to meet. You convert that information into POs as a processor and provide the POs' data to the co-manufacturer for receiving."
  Applying this concept, SIPOC shows a process's supplier, the inputs received from them, the process, and the outputs of that process in a high-level map.
- **Flowchart**: "I'm familiar with those diagrams, with different symbols and shades," Sarah says.

Flowcart Symbols

Flowline	Terminator	Process	Comment
Decision	Stored data	"or" symbol	Input/output
Display	Document	Delay	Manual input
Manual operation	Off-page connector	On-page connector	Summoning junction symbol
Alternate process	Predefined process	Multiple documents	Preparation or initialization

"I have experience in all of that too. It takes forever to complete a process map. That's great in theory and at school," says Jon, while looking at Maria and Tim, "but not here."

"I agree 100% that process mapping can be challenging because it involves cross-functional efforts. With the right practical approach, process mapping is worthwhile. We can do better than competitors, right Mike, Jon?" asks Maria. Although there is no verbal answer, Mike nods and Jon smirks.

Flowcharts are diagrams that show the steps, sequences, and decisions of a process or workflow. Like the other types of process maps, they are helpful for visualizing, documenting, and improving processes. There are over 30 standardized symbols that teams can use in creating a flowchart.

A nice diagram from Asana[6]—replicated below—indicates that there are 5 types of flowcharts:
 1. Decision: to make or justify a step
 2. Logic: to uncover and prevent issues
 3. System: to represent data flow

4. Product: to visualize the product creation process
5. Process: to display process outcomes

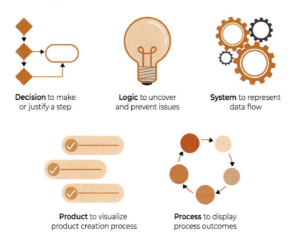

Source: Asana

- **Value Stream Map:** "Wait! Another type! How come? It's getting absurdly confusing!" says Mike.

"I feel the same way," says Tim. "Maybe if you could explain the differences between flowchart, process map, and value stream map it would be helpful."

"Of course!" Maria answers.

A flowchart is a graphic representation that shows the different steps of a process by using symbols. Boxes indicate steps, and arrows indicate how steps connect. A process map is a flowchart with inputs and outputs. A value stream map is a flowchart that shows the time by each process step and the delay between process steps, including transportation. Value stream mapping also indicates whether a step is value-added or non-value-added.

A value stream is a group of activities or steps that need to happen to deliver the product to the customer. Value stream mapping (VSM) shows everything: such steps and steps taken that the customer doesn't value or doesn't care about. VSM allows for a structured visualization of the key steps and data to aid with improvements and optimization of all the processes.

VSM goes further than flowcharts and process maps because they show where the process steps are that create value for the customers, making it easy to know what steps the team needs to eliminate. A value stream map has 3 sections:
1. Information flow—This means the documents and data in the process
2. Product flow—This is the physical flow, including tasks (blue boxes), the person or team performing the action, and key process data. Examples of key process data are cycle time and set-up time.
3. Time ladder or lead time ladder—This is a visual representation of the timeline. The upper portion of the time ladder indicates the average amount of time that a product spends in queue or waiting, while the lower portion of the ladder shows the average amount of time that a person or team works on the product or adds value to it in that stage.

The value stream map is an area of your organization. If there are multiple manufacturing locations, suppliers, and customers, companies create an extended-level map. An extended level map shows the value streams at 60,000 ft., the plant level map at 30,000 ft., and the process level map at 10,000 ft.

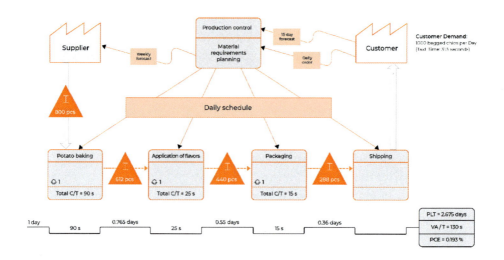

Source: Adapted from Asana

50,000 ft. and 5,000 ft. Process Variables Map

"Our focus is on the process variables map," says Maria. "It also has different views—50,000 ft. and 5,000 ft. She asks the team members to open the six sigma template on the resources page at https://usmsupplychain.com/resources/ that contains the process map, C&E, and FMEA in different tabs.

"Guys, here it comes again, another Chips for Thought!" says Maria.

Chips for Thought

1. Complete the high-level or the 50,000 ft. process map in teams with two or three members. "You gotta be kidding, Maria! How are we gonna do that?" complains Jon.
   "With these instructions," says Maria, waving printouts with her right hand.
2. Present your 50,000 ft. process map results—"Visuals help. Crunchy, flavored, and salty potato and tortilla chips make it an experience!" says Maria.
   "If that works, let's bring a full pallet!" says Ruben.
3. Complete the 5,000 ft. process map in teams with a couple or three members, like you did the high-level process map. There's a tab for that!
4. Present your 5,000 ft. process map results.

Instructions to Complete the 50K ft. and 5K ft. Process Maps in 4 Steps
Step 1—50,000 ft. Process Map
- Identify the process you are mapping, i.e., on-time deliveries
  Use one process step box.
  State the process as a verb. Example: increase on-time deliveries.
- Identify high-level external inputs.

Raw materials
System
People
Incoming information—for instance, customer sales orders, product specs or requirements, must have ready date, payment terms, etc.
- Identify high-level end-customer requirements (outputs)
Include key business measures, if appropriate.
Your project's key "y" should appear as a high-level output. This means that for the project example, on-time deliveries must be listed as an output. "The why for this is coming soon with the C&E matrix." Maria gives a preview. Other outputs are completed, pending, or canceled sales orders.

Step 2—5,000 ft. Process Map—Identify All Steps in the Process
- Include all-value-added and non-value-added steps. This is not the moment to make changes or redesign.
- Keep it simple. Four to six steps are ideal, eight to twelve max.
- "Remember to use verbs because process steps are actions. Lights, Camera, Action!" says Maria.

Step 3—5,000 ft. Process Map—List Key Output Variables
- List key output variables, including process deliverables (actual output) and business measures (consequence of running the process). "Where does on-time delivery go?" asks Maria.

"Business measures," replies Tim, fast but not furious.

"You've got this!" says Maria, "remember that our outputs must be measurable and specific."

Step 4—List Key Input Variables, and Classify
- List all input variables and classify them into controllable inputs with a "C" and uncontrollable inputs ("U" or "N").
- Like outputs, inputs must be measurable and specific.
- "Can an output be the input of another step in the process?" asks Maria.

"No," says Tim.
"Yes," says Amy.

"We have yes and no. What's the right answer?" asks Maria. "What do the rest of the team think?" There are more yesses, and Maria confirms that the outputs of a step can also be inputs of other steps.

"Voila, team, you have your 50,000 ft. and 5,000 ft. process map!" encourages Maria.

C&E

The Cause and Effects Matrix—C&E—is an awesome tool for defining where the biggest impacts and opportunities are. It is like knowing where the big fish are instead of throwing a fishing pole in several ponds.

The funneling effect starts with the C&E matrix after the process variable maps. This funneling effect allows companies to focus on what moves the needle, on what resolves the problem, or y. It's all about the impact. This approach is the opposite of throwing spaghetti at the wall to see what sticks. It's focused and intentional to drive results.

"There's a tab for that in the six sigma template," says Maria. C&E takes all the inputs from the 5,000 ft. process map to include them in the vertical axis of the matrix and the outputs (customer requirements) on the horizontal axis.

Cause and Effect Matrix

		Rating of Importance to Customer											
			1	2	3	4	5	6	7	8	9	10	
	Process Step	Process Inputs	Output 1	Output 2	Output 3	Output 4	Output 5	Output 6	Output 7	Output 8	Output 9	Output 10	Total
1													
2													
3													
4													
5													
6													
7													
8													
9													
10													

Enter C&E Break Point: ___ [Sort & Move To FMEA] [Sort C & E Matrix] [BACK]

"What are we gonna do with this frickin' template or matrix or whatever? It's damn confusing," says Jon.

"The C&E matrix can look intimidating when projects are complex, like here at Alex's Snacks," replies Maria. "When we go through the C&E, we identify the variables with the highest impact and work on them only. We leave the rest of the variables or inputs behind."

Reassuming the example with on-time deliveries, the customer-required date can be an x or input. The question would be, "What's the impact of the customer-required date on on-time delivery?" Unequivocally high. The team assigns a high score. Another input can be payment terms with the customer. In this case, the question would be, "What's the impact of customer payment terms on on-time delivery?" Probably low, corresponding with a low score, too.

If there is more than one outcome, the team needs to rank each outcome based on its importance to the customer. For example, another outcome in the project example could be the lead time to complete the customer order or in-full deliveries. "Remember our friend, the SCOR model with the metrics hierarchy?" asks Maria.

In the C&E matrix, the team scores the key outputs based on their importance to the customer. When teams ask the question, "What's the impact of INPUT on OUTPUT?" they can add the specific process step per the 5,000 ft. map. The complete question to facilitate the scoring of each input is as follows:

What's the impact of INPUT on OUTPUT in the PROCESS STEP?

"In plain English, please!" asks Mike.

"You've got it, Mike," replies Maria. "What's the impact of the customer-required date (INPUT) on on-time deliveries (OUTPUT) in receiving the customer sales order (PROCESS STEP)?"

"Another Chips for Thought, Maria?" asks Ruben.

"You bet," replies Maria, "execution is vital. We cannot live or create results without it."

Chips for Thought

1. Complete the C&E matrix in teams with two or three members by using the template provided on the resources page at https://usmsupplychain.com/resources/; (it's easier, but if you choose to, you can do the manual calculations). "Relax, we have step-by-step instructions," says Maria.
2. Analyze the results and get ready to move on to the FMEA.

Instructions to Complete the Cause & Effects Matrix (C&E) in 5 Steps

Step 1—Identify key customer requirements.
- Identify the outputs from the 5K ft. process map. In the project example, on-time deliveries are the main outcome. Other potential outcomes are in-full deliveries and customer order completion lead time.

Step 2—Assign priority to outputs.
- On a scale from 1 to 10, with 10 being the highest, assign a priority factor to each outcome. In the on-time delivery project, on-time delivery has a 10 as a priority factor. In-full deliveries and order completion lead time have lower priority factors.
- Rank the outcomes based on priority factors. In the example, on-time deliveries rank first.

Step 3—Identify all process steps and inputs.
- From the 5,000 ft. process map, identify all process steps—boxes in orange—and all inputs.

Step 4—Score each input based on impact on output
- Evaluate the correlation of each input to each output. "Correlation? What?!" says Jon.
  "Correlation is the question: What's the impact of INPUT on OUTPUT in the PROCESS STEP?" answers Maria.
  "Jon, it's the example with 'What's the impact of the customer-required date (INPUT) on on-time deliveries (OUTPUT) in receiving customer sales orders (PROCESS STEP)?" replies Mike, looking at his workbook.
  "Understood, a fancy term from consultants," says Jon.
  Maria smiles.
- When the correlation is high, the score is high. This means that the impact of the input variable is significant on the output variable (y). In other words, changes in the input variable are like a typhoon for the output variable.
- When the correlation is low, the score is low. The impact of the input variable is low on the output variable. Changes in the input variable are like a baby wave for the output variable.

Process Mapping, Matrices, Frameworks, and Toolkit from Lean Six Sigma 115

Step 5—Get the weighted-average scores.
- The template does it for you. It cross-multiplies the correlation values or impact on the outputs (y's) with priority factors—this is the importance of customer requirements—and sums it all up.
- By analyzing the results, the team can focus on the input variables that have the highest impact on the most pressing customer requirements. The input variables with the highest score continue their journey to the FMEA matrix while the team says goodbye to those with low impact.

FMEA

"Good morning Ernesto, we are about to cover the FMEA matrix—the quintessential six sigma tool in Villanova University words.[7] Are you available to join us?" asks Maria.

"Geez! I'm going to be traveling all this week…I need to be on-site at the co-manufacturer for baked cheese curls. We're having production issues and have large customer orders to deliver. If we can't make it on time, we get in trouble because our customers will be upset," says Ernesto to Maria in a Teams call.

"On-time deliveries is the project we're using as an example for the lean six sigma tools," says Maria.

"Really? I wish I could be there," says Ernesto.

"No need to be here in person, Ernesto," explains Maria, "we can cover FMEA online. Even before Covid, companies have used six sigma tools online with global teams. This is not an issue at all."

"Perfect! Please send me the information to join you, guys."

During the FMEA training, Maria explains that the FMEA is a structured approach to identify ways the product or process can fail and eliminate or reduce the risk of failure. "This matrix is in your six sigma template, on the resources page at https://usmsupplychain.com/resources/," Maria says.

"Good, there's a tab for that, too, Maria," says Ruben.

Process/Product: Failure Modes and Effect Analysis (FMEA)

Process Step	Key Process Input	Failure Modes - What can go wrong?	Effects	SEV	Causes	OCC	Current Controls	DET	RPN	Actions Recommended	Resp.	Actions Taken	SEV	OCC	DET	RPN

"Yes, a tab for that," Maria reiterates, "and we have more funneling effect with each tab. After the FMEA, there are between 8 and 10 variables left. The starting point is the input variables that we carry from the C&E matrix."

With the FMEA, companies can do the following in a structured approach:
1. Identify what could go wrong with a product or process. For the on-time deliveries project example, Ernesto mentions production issues. These production issues can be a breakdown of machines, raw materials or packaging that doesn't meet the specs (for example, the film rolls are too big for the machine to run) or that arrive late (the marine salt gets there two weeks late).
2. Estimate the risk or impact associated with specific causes.
3. Prioritize actions to reduce the risk.
   "What are the sources of risk?" Maria asks the team.
   "So many," answers Ernesto, "what do you think, team?"
   "Machines, QC issues, raw material variation, to name a few," says Mike.
   "There's always more, right? I feel like issues are never-ending," says Ernesto.

"Right, but not all of them have the same impact," says Maria. "If you look at the template you can see these columns."

- **Failure mode**
  Formal definition: The way a specific input fails.
  Example with on-time deliveries: the packaging supplier delivers the film late.
- **Effect**
  Formal definition: Impact on customer requirements.
  Example with on-time deliveries: late order delivered to the customer.
- **Cause**
  Formal definition: Sources of process variation that cause the failure mode to occur.
  Example with on-time deliveries: The packaging supplier gets their raw materials late. "We want to start identifying causes for the failure modes (what could go wrong) with the highest severity rankings," Maria says. "Agreed. We want to focus on the right things," Ernesto adds.
- **Current control**
  Formal definition: Systems and methods in place to prevent or detect failure modes or causes (before causing effects).
  Example with on-time deliveries: Weekly calls with the packaging supplier to check on schedule and potential issues.
- **Severity (of Effect)**
  Scale: 1=Not Severe, 10=Very Severe
  Formal definition: Importance of effect on customer requirements.
  Example with on-time deliveries: 10 if the delay with the packaging supplier prevents the team from delivering to the customer on time. "Can we give it a 400, 500 instead of a 10?" Ernesto jokes.
- **Occurrence (of Cause)**
  Scale: 1=Not Likely, 10=Very Likely
  Formal definition: Frequency of the cause.
  Example with on-time deliveries: 8 or 9 or 10 if it is common that the packaging supplier gets their raw materials late.
- **Detection (Capability of Current Controls)**
  Scale: 1=Likely to detect, 10=Not Likely at all to detect
  Formal definition: Ability of systems, methods, and tools to detect or prevent the causes or the failure mode.
  Example with on-time deliveries: 4 or lower if the packaging supplier weekly calls are effective and allow the team to take action. Greater than

4 if delays come out of the blue (or out of any other color too without due notice).
- **Risk Priority Number (RPN)**
  Formal definition: Severity X Occurrence X Detection
  Example with on-time deliveries: if severity = 10; occurrence = 8; detection = 6; the RPN = 480

"This is the first Chips for Thought for which Ernesto has joined us," says Maria.

"We need him because I'm sure it is on the FMEA," says Tim.

"Spot on, Tim. This Chips for Thought is on the FMEA and builds on our C&E. Can anyone brief Ernesto so the full team is aligned?" asks Maria.

"I can do it," says Amy, raising her hand.

"We defined the problem—increase on-time deliveries in a way that we can measure. The problem is the y. Then we created process maps with process steps and all the x's that are the inputs." Amy starts explaining by sharing her screen with Ernesto.

"After that, we took all the inputs and prioritized those with the highest impact on on-time deliveries. We ranked them and moved the top ones to the FMEA." Amy concludes her explanation.

"You've passed with flying colors," Maria celebrates.

"May I add something else?" asks Sarah.

"Of course!" replies Maria.

"About the funneling effect…in the process map we had every possible input (x's), in the C&E fewer x's, and with the FMEA, we expect to have fewer," Sarah states.

"Yeah, about 8 – 10," adds Tim.

"Got you, so we can focus on those," reinforces Ernesto.

"I think we are ready, team," says Maria.

Chips for Thought

1. Complete the FMEA matrix. "This time everybody is on the same team," Maria clarifies.

"Are there any instructions? How are we gonna do this?" asks Jon.

"We have here step-by-step instructions. Any other questions before the fun begins?" asks Maria.

"Ready, set, FMEA!" says Maria.

Instructions to Complete the Failure Mode & Effects Analysis Matrix (FMEA) in 9 Steps

Step 1—Determine failure modes.
- For each process input, identify what could go wrong. These are the failure modes.

Step 2—Determine effects of failures on customers.
- For each failure mode, determine the effects of failures on customers. "I don't even want to think what's going to happen if we are late with the orders on cheese curls," says Ernesto, hiding his head between his arms.

Step 3—Identify potential causes.
- Explore potential cause(s) of each failure mode. In the project example, a cause is the delay that the packaging supplier experiences with their

raw material suppliers, Tier 2 suppliers for Alex's Snacks. There could be other causes like tight schedules or capacity challenges at the packaging supplier.

Step 4—Document current controls
- List current controls for each cause or failure mode. In the on-time deliveries project, there are weekly calls with the packaging supplier. This is a current control in place. Another control can be the order confirmation—including the have-ready date—on the packaging supplier's end.

Step 5 – Define rating scales for Severity, Occurrence, and Detection.
- You may use the scale included below.

FMEA RPN Scale

Severity		Occurrence		Detection	
Hazardous without warning	10	Very high: Failure is almost inevitable	10	Can not detect	
Hazardous with warning	9		9	Very remote chance of detection	
Loss of primary function	8	High: Repeated failures	8	Remote chance of detection	
Reduced primary function performance	7		7	Very low chance of detection	
Loss of secondary function	6	Moderate: Occasional failures	6	Low chance of detection	
Reduced secondary function performance	5		5	Moderate chance of detection	
Minor defect noticed by most customers	4		4	Moderately high chance of detection	
Minor defect noticed by some customers	3	Low: Relatively few failures	3	High chance of detection	
Minor defect noticed by discriminating customers	2		2	Very high chance of detection	
No effect	1	Remote: Failure is unlikely	1	Almost certain detection	

Step 6—Assign ratings per scale.
- Assign severity, occurrence, and detection ratings to each cause.

Step 7—Calculate RPNs.
- For each cause, calculate RPNs based on the formula: Severity X Occurrence X Detection

Step 8—Define recommended actions.
- Suggest actions to reduce high RPNs. An example is to increase the frequency of calls with the packaging supplier or involve a Tier 2 supplier in the call as well to facilitate detection.

Step 9—Complete recommended actions.
- Recalculate RPNs after following the recommended actions. After implementing the recommended actions, reassess severity, occurrence, and detection. In the example, the detection rating becomes lower, leading to a lower RPN.

FTA – Fault Tree Analysis

The FTA (Fault Tree Analysis) is an approach to analyze the failure causes from top event to basic characteristics to identify the root causes and find the failure mechanism. "You may think I'm a gardener—strategic goal tree, fault tree analysis, the supply chain tree...I've just made up the last one," says Maria, while the team members laugh, "but this tool is simple and helpful for root cause analysis."

There are three key components in an FTA:
1. The diagram of the process—This is a flowchart. The analysis consists of drawing logical conclusions that start with the failure event and progress through each subsequent event until finding the root cause.
2. The events that have occurred—An event is an occurrence in the system. There are input events that cascade into other events and output events that are results of input events. "These are the FTA symbols," says Maria, pointing at the table below.

FTA Symbols

Symbol	Name	Meaning
	And gate	Event above happens only if all events below happen.
	Or gate	Event above happens if one or more of events below are met.
	Inhibit gate	Event above happens if event below happens and conditions described in oval happen.
	Combination gate	Event that results from combination of events passing through gate below it.
	Basic event	Event that does not have any contributory events.
	Undeveloped basic event	Event that does have contributory events, but which are not shown.
	Remote basic event	Event that does have contributory events, but which are shown in another diagram.
	Transferred event	A link to another diagram or to another part of the same diagram.
	Switch	Used to include or exclude other parts of the diagram which may or may not apply in specific situations.

3. The connections between the events—gates—link events with "and" or with "or." In the example with the line shutdown, if there is a delay with the packaging and also with the ingredients, the gate to connect these two events is "and." If the delay is with the packaging only, the gate to connect packaging materials and ingredients is "or."

Although both the FMEA and FTA apply to root cause analysis, they are different. The key difference lies in the how. FMEA is a "bottom up" approach, looking at each input variable and building a list of all potential failure modes. The FMEA is a kind of Murphy's law where anything that can go wrong will go wrong to build the list of failure modes. FTA, instead, takes a "top down" approach, starting with the failure and then diagnosing what could have caused the problem with questions.

"How do both approaches work with our on-time deliveries project?" asks Ernesto.

"With the FMEA, we have on-time deliveries as an output and we analyze all the input variables coming from the C&E matrix, such as raw materials, packaging, co-manufacturer's production capacity, co-manufacturer's production schedule, etc. We think about every possible aspect that could go wrong and assess severity, occurrence, and detection. With FTA, we start with what goes wrong—the production issues with cheese curls—and go backwards to identify the cause. It is a 6-step process.[8]" Maria explains.

The 6-step process of the FTA is as follows:

Step 1: Define the top event, the failure. This is the go point for the diagram. The more specific the top event, the better results with the FTA. "Production issues with cheese curls is too broad," notes Maria.

"You're saying that the boss is wrong, Maria? How daring you are!" says Ruben in a goofy fashion.

"I need to clearly define what I mean by production issues," Ernesto shares, as if he is learning a lesson from Mr. Miyagi (Pat Morita) in Karate Kid, an American martial arts drama movie. After all, lean six sigma also uses karate terminology.

Step 2: Understand the system and how the different moving pieces work together. The goal is to identify how the system or process was working before the failure or top event.

"How was the process working before the line shutdown screwed up everything?" asks Mike, candid as a three-year-old.

"That's right, Mike," confirms Maria.

Step 3: List potential causes of the top event. There are different ways to accomplish such a list. This is a have it your way like Burger King promotes their products or "My Way" by Frank Sinatra. One potential way is as follows:
- Identify five potential causes.
- Estimate the probability of each being the cause of the top event.
- Rank the causes based on probability.

Step 4: Draw the fault tree diagram.

No need to be a DaVinci or Van Gogh to draw the fault tree diagram. Teams start with the top event; from there, they map out the potential causes and connect the steps with the "and" and "or" gates.

Step 5: Assess risk.

For risk assessment, companies assign risk and probability to each base event.

Step 6: Mitigate risk.

This last step is about taking action to mitigate the events with the highest risk and probability.

"That sounds similar to the RPNs in the FMEA," says Amy.

"So, when do we use the FMEA and when the FTA? It's a lot of work, so we need to pick the right one," says Jon.

"Understood. Let's see the differences between these two tools and when to apply the FMEA or the FTA," says Maria.

Use FMEA when:[9]
1. It's challenging to identify a top-down failure event to begin an FTA.
2. It's important to identify every possible failure mode, even those with low severity or low occurrence.
3. It's simple to predict all failure modes.

Use FTA when:
1. It's straightforward to identify a top-down failure event.
2. The process or system is complex, with strong interaction between the different moving pieces.
3. There's room for human error or external factors to cause issues.

A Fault Tree Analysis Example

```
                  Mixer stopped
                    working
                        |
          ┌─────────────┴─────────────┐
   Ingredients late              Packaging late
      deliveries                    deliveries
                                        |
                              ┌─────────┴─────────┐
                          Film late         Corrugated boxes
                          deliveries        late deliveries
                                                  |
                                              Overload
```

5 Whys—the "Toddler Approach"

Another method for root cause analysis is the 5 Whys or 5Y. Like toddlers asking "why?" many times over, the team asks "why?" 5 times to uncover the root causes, to peel the onion or the potato at Alex's Snacks. It's effective in getting under band-aids to identify the root cause without investing significant time or resources.

The core or the salt on the chip is asking "why?" five times to get to the root cause. Below are the 7 steps to conduct the 5 Whys, the "toddler approach":

1. Assemble the right team—it's critical to have the right expertise and experience together to answer the 5 why questions.
2. Identify the problem—"Picking on Ernesto again, we want to be specific in defining the problem with the 5 Whys as with the FTA," says Maria. "You're getting into trouble, Maria," warns Ruben jokingly.
3. Determine why the problem took place. This is the first why to get a cause and then we dig deeper or peel the potato.

4. Ask why again. This is the second why to advance further in the path to the root cause.
5. Repeat. The team keeps asking why until getting to the root cause. On average, it takes 5 whys.
6. Execute. This step is about taking fast action to implement measures to prevent the problem from happening again.
7. Track the results and adapt or pivot.

"Five Whys seems a piece of cake relative to the FMEA and the FTA," says Mike.

"Can we become toddlers and do 5 Whys only?" asks Ruben.

"I like 5 Whys too, but it has its limitations. Unfortunately, it's not a good fit for complex situations, as it considers only a string of causes and effects—there's no branching out." Maria says. "To address this, some companies use three-legged Five Whys that allows branching out."

Maria wraps up the Five Whys with, "It works best for root cause analysis of simple, one-dimensional problems."

"Great progress, team! We've gotten a lot done with the FMEA and FTA. There's more to do, but we are moving forward," Ernesto says and adds, "I will reconnect with you all later. Thank you so much, guys!"

MSA

The team advances to the MSA. Lean six sigma involves data, statistics, and measurements to eliminate waste (lean) and to reduce variation (six sigma). In this context, it is imperative that companies define a measurement system that they can trust; otherwise, they can get misleading results. It could happen that the team reports an improvement in on-time deliveries to Ernesto because of a failure in the measurement system. Here is where the MSA comes into play.

The MSA or measurement system analysis is a statistical study that defines whether the measurement system is providing reliable data for decision-making. A Continuous Gage R&R study is a statistical study for continuous data, and Attribute Gage R&R for discrete data.

Continuous data has an infinite number of possible values, like the weight of Alex's Snacks products. In the potato chips and tortilla bags you can read 4.4 oz., 10 oz., 18 oz., etc. Discrete data includes non-divisible figures and statistics you can count. For example, the bags count in a case or the number of cases in a pallet.

In measuring an object, the value that the measurement system provides is the sum of the natural variation of the object and the variation from the measurement system. When Alex's Snacks weigh its products, the number displayed on the scale is the sum of the variation in the product's weight plus the variation in the mechanism of the scale. An MSA study allows the team to know about the variation from the scale. If it is low, great news; if it is high, the team needs to replace the scale.

In an MSA study, there are six aspects to look at:
1. **Repeatability**—it is repeatable if the same person measures the same object with the same device multiple times and gets the same result. "For example, if Jon measures today the on-time deliveries in August and gets the same results as those he got yesterday," clarifies Maria.
2. **Reproducibility**—It is reproducible if multiple people measure the same object multiple times with the same device and all get the same results. "If Tim, Sarah, Mike, and Amy measure the on-time deliveries in August, all get the same results as Jon," Maria indicates.
3. **Stability**—It is stable if the variation holds steady over time. "We get the same variation week after week, for instance," Maria points out.
4. **Bias**—It is a one-directional tendency of the measurement system. "For instance, let's say that we are measuring on-time deliveries based on a supplier's report. The report considers Saturday as a valid delivery date although the customer cannot use the shipment until Monday," states Maria.
"Or when the delivery is on Friday at 5 pm, right?" asks Mike.
"Exactly. In those situations, we need to adjust the MSA," says Maria.
5. **Linearity**—It is linear if the measurement system can hold steady over the continuous spectrum of the desired measurements. "We want our

measurement system to work for a few shipments and for lots of them too," explains Maria.

6. **Discrimination or resolution**—This is the capability of producing meaningful values. "Do we want to know the on-time deliveries based on weeks, days, or hours? What is good enough? These are the questions that we want to answer," says Maria.

"Eager to see the toolkit for the Analyze phase?" asks Maria.

"Let's do it," says Tim.

Analyze

In the Analyze phase, companies may apply root cause analysis tools included in the Measure phase such as FMEA, FTA, and the 5 Whys. Teams can also apply more than a tool. For example, they can create a fishbone diagram and then follow with the 5 Whys.

Ishikawa Diagram—Fishbone Diagram—Cause and Effect

"I bet you have heard about this tool before, maybe as the fishbone diagram?" asks Maria.

"Yes, I have," confirms Sarah.

"Me too," say Mike and Tim, while Jon remains silent.

The Ishikawa or fishbone diagram is a visual tool for root cause analysis. The completed diagram looks like a fish. It can look like a small pond fish or a humongous shark with the different branches.

The head of the fish—no matter the size—is the problem statement, the y. In the project example the y is the on-time deliveries we want to increase. There is a horizontal line from the head with branches (here the trees again) or bones to show potential causes.

Many teams use the 6Ms as the bones categories:
1. Man (staff)
2. Machine
3. Methods
4. Materials
5. Measurements
6. Mother Nature (environment)

There are two main ways to build the fishbone diagram:
1. Start from the problem and brainstorm potential causes in each category.
2. Start with the potential causes and put them into the categories. "We use sticky notes with this approach, either in person or online, one cause per sticky note," indicates Maria.

The cause-and-effect diagram is not quantitative. It applies when there is no hard data or as a preliminary work to then use other tools from the toolkit. Once the team identifies the root cause, take actions to address it. Crush it, baby!

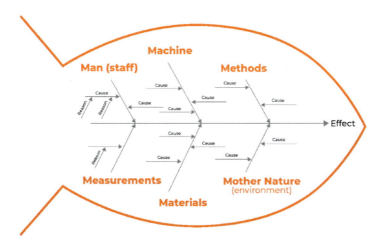

Scatter Graphs
Scatter graphs, diagrams or plots are tools to show whether there is a relationship between two sets of data and provide a visual correlation coefficient. The goal is to identify patterns by plotting the data points.

The resulting scatter plots show correlation only. They do not prove causation because there could be another factor. "There is a typical example to show that correlation does not imply causality," Maria starts explaining. "It's about sharks and ice cream sales. The scatter graph may show a correlation between the two, but ice cream doesn't cause shark attacks. There's another factor—heat—that drives both shark attacks and ice cream."

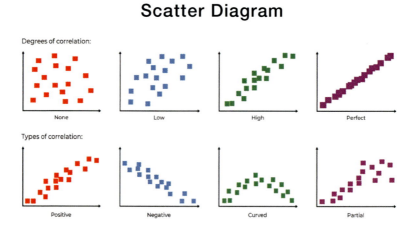

Pareto Charts

The Pareto chart allows for visualization of the 80-20 rule. In the project example of increasing on-time deliveries, 20% of the causes represent 80% of the impact on the problem, on the y.

The Pareto chart is a combination of a bar graph and a line graph. The horizontal axis displays the potential causes in descending order based on frequency (y-axis). Each bar represents a cause. The line graph shows the cumulative percentage of the impact of the causes on the y, from 0 to 100%.

In the project example, delays with packaging or with raw materials, co-manufacturer's schedule constraints, or quality issues could all be potential causes with different frequencies of occurrence. The team includes each cause on the x-axis starting with the most frequent causes. The line graph adds up the frequency of the different causes from 0 to 100%.

Pareto Chart of Late Customer Deliveries

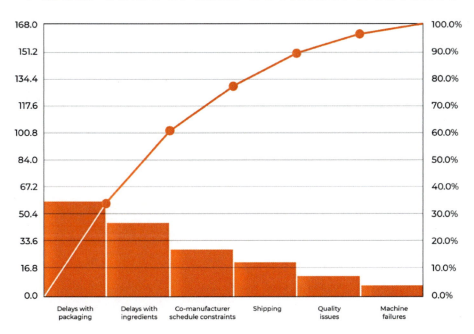

Non-Value-Added Analysis

"Is this related to waste? To what does not add value to the customer?" asks Amy.

"It is!" reinforces Maria.

The focus on non-value-added analysis is identifying which process steps add value to the customer and which do not. Value means what the customer is willing to pay for. When companies identify non-value-added activities, they direct their efforts to eliminate them.

t-Test

"Oh no, Maria! t-test? You scare the scrap out of me with that name," says Ruben.

"No worries, Ruben, team. When it comes to statistical tools, I want to show you what's out there available to use, and then we go deeper if needed."

"Thank goodness!" says Ruben, pretending to wipe away forehead sweat.

Teams use a t-test to evaluate a statistical parameter—the mean, for instance—of one or two populations using hypothesis testing. "Let's take the weight of the potato chip bags filled with product, as the variable needs to be continuous and have a normal distribution to run the t-test," Maria explains. "We can analyze if the mean of a sample is significantly different than the target or set value of 4.4 oz. This is a one-sample t-test."

"How many potato chip bags are in the sample?" asks Tim.

"Over 30 would be fine because of the Central Limit Theorem," replies Maria.

"Okay, another tequila shot for me," says Ruben.

After laughing, Maria explains that per this theorem, if the number of observations in the sample is around 30, we can assume that the sample means follow a normal distribution.

A two-sample t-test can be used to compare the weight of the filled chip bags from two different co-manufacturers (two different populations). The null hypothesis would be that the weight of filled chip bag samples does not differ between co-manufacturers. The alternative hypothesis would be that they differ significantly.

There are other types of t-tests, like the paired sample t-test (for example, to compare "before" and "after") in the same population, the equal variance t-test, and the unequal variance t-test.

Before collecting the data for any kind of t-test, teams should decide whether they will have a one-tailed or two-tailed test. For instance, if it's enough to know that the weight of the chip bags between the two co-manufacturers is different, teams perform a two-tailed t-test[10]. If, instead, the teams want to know which weight is greater between the two co-manufacturers, the one-tailed t-test is helpful.

"How about a step-by-step process to perform a t-test?" asks Maria.

"Yes, please," answers Sarah.

How to Perform a T-test Step-by-Step[11].
1. Define null (H_0) and alternative (H_a) hypotheses before any data collection.
2. Define a significance level alpha (α). Alpha is the risk that you are willing to take of reaching a wrong conclusion. For example, if α=.05, it means that you are taking a 5% risk of drawing the conclusion that the weight of the chip bags of the co-manufacturers is different when they are not. More risk-averse persons can set α=.01 (1% risk).
3. Check data.
4. Check the assumptions for the test.
5. Conduct the test and get your conclusion. Performing a t-test for means involves calculating a test statistic (t-value).

Statistical software Minitab[12] indicates that "a test statistic is a standardized value that is calculated from sample data during a hypothesis test." The procedure then compares the data against the null hypothesis, considering sample size and variability in the data.

Each t-test performed provides a t-value only. With multiple t-tests and t-values, the team can plot them to obtain the sampling distribution. A t-distribution has degrees of freedom that is a value related to the sample size. By placing the t-value from the study in a known t-distribution, the team can see how consistent the results are with the null hypothesis and calculate the probability of that t-value.

A probability indicates how common such a t-value is under the assumption that the null hypothesis is true. "If we talk about probability, we talk about p-values," states Maria.

"I have heard about p-values before!" exclaims Tim.

"We use p-values across the board in statistics," Maria says, "as p-values indicate whether there is statistical significance in a hypothesis test."

Based on an article by Minitab[13], "in technical terms, a p-value is the probability of obtaining an effect at least as extreme as the one in your sample data, assuming the truth of the null hypothesis. This means that p-values measure how compatible the data is with the null hypothesis. They don't measure support for the alternative hypothesis. "In our example, p-values show how compatible the data is with the mean of the weight of the filled potato chip bags not being significantly different between the two co-manufacturers," clarifies Maria.

High p-values: the data is likely with a true null hypothesis.

Low p-values: the data is unlikely with a true null hypothesis.

"We are assessing the possibility of using one more co-manufacturer for the production of tortilla chips. Can we use a t-test with three co-manufacturers?" asks Mike.

"Nope," Maria answers, "we can use the t-test with up to two populations or groups. If we have three co-manufacturers or want to make multiple pairwise comparisons, we need to use ANOVA."

ANOVA

Teams apply ANOVA (Analysis of Variance) to determine if the mean of three or more groups differ significantly. A one-way ANOVA[14] considers three sources of variability:
1. Total—It considers all variability between observations.
2. Between—It refers to the variation between subgroup means.
3. Within—It relates to the random variation within each subgroup (noise).

"We have one-way ANOVA and two-way ANOVA," says Maria, "and for the situation with the three co-manufacturers that Mike suggests, we can apply one-way ANOVA." One-way ANOVA considers one element—in our example, weight—and measures whether the mean of the alternatives—in our example, weight by each of the three co-manufacturers—is significantly different.

The two-way ANOVA adds another factor that impacts the output variable. In the example we are following, the additional factor could be type of product—potato chips vs. tortilla chips—in addition to the three co-manufacturers to understand their impact on the weight of the filled chip bags.

"All of this sounds wonderful," says Jon, "but are we working on the weight or on the on-time deliveries?"

"I hear you, Jon, loud and clear," states Maria. "We take the example of the weight because t-test and ANOVA apply when the dependent variable is continuous. Let's see now the Chi-Squared Test that applies to discrete variables like on-time deliveries."

Chi-Squared Test

The chi-squared test is for hypothesis testing when both the input and output variables are discrete. There are a couple of common chi-squared tests:
1. Chi-square goodness of fit test
2. Chi-square test of independence

	Chi-Square Goodness of Fit Test	Chi-Square Goodness of Independence
Number of variables	One	Two
Purpose of Test	Decide if one variable is likely to come from a given distribution or not	Decide if two variables are related or not.
Example	Decide if packaging supplier delivers film on-time to both co-mans in the same proportion.	Decide if the customer orders that Alex's Snacks delivers late are related to the co-man the company uses.
Hypothesis	H0: Proportion of packaging on-time deliveries to both co-mans is the same. Ha Propotion of packaging on-time deliveries to both co-mans is not the same.	H0: Proportion of customer orders delivered late is independent of the co-man used. Ha: Proportion of customer orders delivered late is different, based on the co-man used.
Theoretical Distribution used in Test	Chi-Square	Chi-Square
Degrees of freedom	Number of categories minus 1. In our exmaple, 2 co-mans minus 1.	Number of categories for first variable minus 1, multiplied by number of categories for second variable minus 1. In our example, 1 (because late deliverie are a Yes/No variable and 2-1=1).

Source: JMP[15]

Design of Experiments (DOE)

Per the American Society of Quality (ASQ)[16], DOE is a branch of applied statistics that deals with planning, conducting, analyzing, and interpreting controlled tests to evaluate the factors that control the value of a parameter or group of parameters. This is different from empirical observations because the team takes the driver's seat by controlling the variables as opposed to being in the passenger seat observing.

DOE allows for the manipulation of several input variables, defining the effect on the outcome. This represents an advantage over experimenting with one factor at a time because the team may miss important interactions. In a full factorial DOE, teams investigate all possible combinations, while in a fractional factorial, only a portion of the scope.

DOE is a powerful tool that applies to the following situations:
- Finding optimal settings
- Screening several factors to determine those with the highest impact
- Identifying the factors with the lowest impact
- Confirming input/output relationships to develop a predictive equation for what-if analysis
- Reducing the time and number of experiments needed to test multiple factors

"Do you have a step-by-step process for DOE?" asks Amy.

"Here it is! Experiment and enjoy, too," says Maria.

Instructions for DOE in 5 steps
1. Identify and get full knowledge of inputs and outputs to include in the investigation.
2. Define appropriate measure for the output (variable is preferable). Flowcharts or process maps are great points to start.
3. Create a design matrix for the factors under the investigation. The design matrix shows all possible combinations of high and low levels for each input factor.
4. Conduct the experiment under the prescribed conditions.
5. Evaluate the results.

Taguchi
The Taguchi method—referred to as a robust design method—is a statistical method like DOE. The main difference between Taguchi and DOE is their approach to the different variable interactions.

DOE assumes that all inputs interact with all other inputs, while Taguchi considers that there is some existing knowledge about the process. This existing knowledge makes Taguchi's method more efficient because the team doesn't conduct tests for those interactions that they already know are inexistent.

Another difference between DOE and Taguchi is that Taguchi methods distinguish between factors that are controllable (control factors) and those that are not (noise factors). By applying Taguchi methods, teams may test each combination more than once at different levels of noise factors. This is possible due to the reduced number of tests.

Teams apply the Taguchi methods[17] when there is a reasonable number of factors—from 3 to 50—few lower-order interactions between factors, and when there a few significant factors from the statistical perspective.

Regression Analysis

"Is regression analysis more numbers and statistics? I'm gonna grab more coffee. Anyone want something?" asks Mike with tired eyes, looking forward to getting his hot beverage.

The rest of the team also takes a short break to recharge.

Regression analysis is a tool that teams apply in the Analyze phase of DMAIC to identify root causes and defects and in lean to identify waste. Regression analysis makes predictions and measures whether the results are consistent with the expected results when variables change.

Together with scatter plots, regression analysis defines a model—a mathematical equation—that shows the impact of the input variables on the outcome variable and predicts future performance. Linear regression implies one input variable and one output variable. Multiple linear regression implies more than an input variable (more than an x).

To predict the value of the outcome variable (y), teams plug the x's' values into the equation. If these values change, the process reiterates by plugging in the updated values to predict the value of the outcome variable (y).

"You can't imagine how hard I'm trying not to say the four-letter word," says Jon, "this is for a PhD or data scientist, but not for us. I can't care less about it. We have the issue with lost sales because of late deliveries and I don't see how this thing can help us. It's a waste of everybody's time."

"Jon, the terminology can get confusing, and it can become challenging to see the connection of these formulas and studies to our project to increase on-time deliveries. My goal is to give some explanations and then get into motion," Maria replies.

"What regression analysis does is give us an equation like this—please note this is an example I've made up to explain:

$$y = 103 + 63x_1 + 13x_2$$

Where y = on-time customer deliveries
x1 = on-time packaging deliveries
x2 = on-time oil deliveries

You plug in the values for x1 and x2 and you get the on-time customer deliveries. Simple, right? We need to see how to get to the actual equation to predict on-time deliveries. Better now, Jon?"

"Better. I think you should have started with the example and then go over the crazy stuff," says Jon.

"Thank you! I'll keep this in mind for future trainings. Thank you very much for the feedback," Maria answers.

Best subsets and stepwise regression[18] suggest models (equations) to use. Stepwise regression selects a model by adding or removing individual predictors, a step at a time, based on their statistical significance. Best subsets compare all possible models from a set of predictors and shows the best-fitting models that contain one predictor, two predictors, etc.

Both stepwise and best subsets build their models on the predictors teams specify. With stepwise regression, teams get one suggested model, and with best subsets, teams get more than a model to select.

"To end the Analyze phase's tools in a high note like Lady Gaga in her song 'Shallow,' let's have a Chips for Thought!" says Maria.

Chips for Thought

1. Select two or three tools from the toolkit in the Analyze phase to apply to your project.
2. What issues have you encountered?
3. What works?
4. What doesn't?
5. Based on the lessons learned, what would you do differently next time?

"Let's get into the toolkit for the Improve phase. Expect awesome tools, fun, and some Japanese words too!" advises Maria.

Improve

The toolkit for the Improve phase includes the following:

Affinity Diagram—"Post-brainstorming With Order"

Japanese anthropologist Kawakita Jiro invented the Affinity Diagram[19]—also referred to as the K-J Method in a generalized way by some websites—which organizes a large number of ideas into their natural relationships. This is the reason Maria refers to the affinity diagram as "post-brainstorming with order."

A great initial point for the Affinity Diagram is the output of a brainstorming session. For a brainstorming session to work, all ideas are welcome. There are no rights or wrongs. What matters is the participation and engagement of the team members. After the many ideas generated—including those crazy ones—an Affinity Diagram helps to organize and consolidate them based on their similarity in meaningful categories.

"When do we use an Affinity Diagram?" asks Tim.

"Perfect question, Tim!" ASQ indicates the following situations:
1. When there's a chaos of facts or ideas (more than a brainstorm of ideas, it is kind of a hurricane).
2. Too-complex issues
3. When group consensus is necessary

Teams conduct an Affinity Diagram after a brainstorming session, a survey, or other data collection method.

"Today is your lucky day, team," says Maria, "on the ASQ website there is an example of an Affinity Diagram from a hospital working on improving on-time deliveries of medications. Is not that *affinity*?" asks Maria.

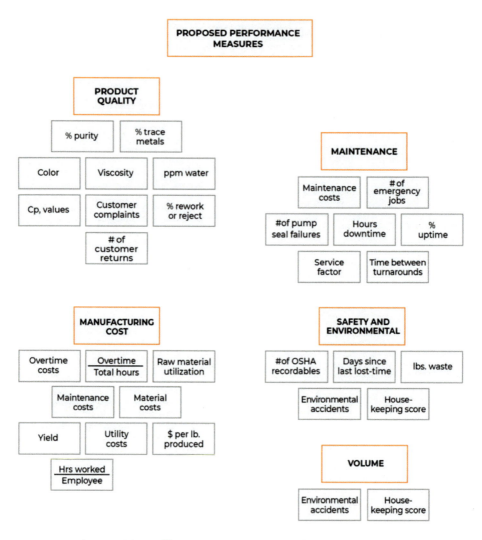

Source: ASQ – Affinity Diagram at a Hospital to improve on-time deliveries of medication.[20]

Maria provides the materials needed and step-by-step instructions to get a flourishing K-J, like the ingredients and instructions to bake an irresistible chocolate cake.

Materials needed: sticky notes or cards, marking pens, and a wall, table or floor to group and move around the sticky notes.

How to Create an Affinity Diagram in 4 Steps

Capture Ideas—One Idea = One Sticky Note
Each team member receives sticky notes or index cards to record their ideas or issues in silence—no collaboration in this step. One idea per one sticky note or card. Each idea needs to be between 3 and 7 words. It's important to use markers to be able to read the sticky notes from any distance.

Group Ideas
Team members look for relationships between individual ideas and sort them into five to ten related groupings. These are groupings with no names or labels. Team members do the sorting in silence without collaboration. It is fine to move the sticky notes back and forth and that some—"lone wolves"—remain without a grouping.

Label Groupings
This is the moment that the team members can talk to each other and define categories for the different groupings. During the discussion, they may decide to move around some of the sticky notes. They may identify duplicates to eliminate. Once everyone is happy with the groupings, the team proceeds with labeling or adding headlines for each of the groups. It is important that the headline—in a different sticky note if it is not already in one of them—be descriptive and captures the essence of the common thread.

Affinity Diagram

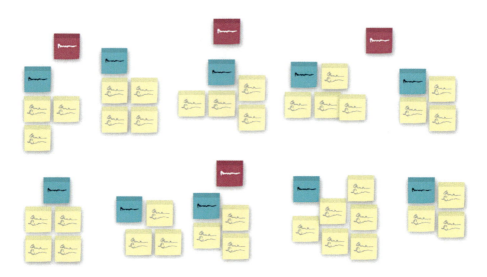

Create Action Plan
Let's make it happen! After the Affinity Diagram is complete, the team creates an action plan. Let's rock and roll!

"Are we gonna have a Chips for Thought on the Affinity Diagram?" asks Amy.

"The Chips for Thought is on K-J, because we have an example of the Affinity Diagram on on-time deliveries on the ASQ website," replies Maria.

"I thought that the Affinity Diagram and the K-J Method are the same," states Tim, opening his eyes wider.

"There are some generalizations that consider that they are the same, but they are not," indicates Maria.

K-J Analysis
The Affinity Diagram and K-J are different[21] in the following aspects:
- **Preparation**—There is little preparation for the Affinity Diagram while the K-J needs a "theme" question to ensure the right data collection.

- **Sources**—The main source in the Affinity Diagram is the team members' ideas. The K-J takes a more data-based approach with data gathering and facts.
- **Grouping**—In the Affinity Diagram, grouping is quick and based on keywords. The teams may create groups with a bunch of sticky notes. There is no limit on the number of sticky notes per group. In the K-J, there's a limit of 3 sticky notes per group. The team thinks about "the story being told" by the data.
- **Headings**—The headings are quick and informal in the Affinity Diagram, while they have structure in the K-J. The team uses complete sentences with a conjugated verb.
- **Reflection**—Many times the result is different categories in an Affinity Diagram. In the K-J, the team works on cause and effect relationships and votes to assign priorities. There is a full final statement that translates the findings.
- **Language**—the K-J uses a more elaborate language than that in the Affinity Diagram. There are different levels of grouping or labels in the K-J that correspond to the "ladder of abstraction." The factual data is in black at the bottom of the ladder. Moving up the ladder, there's the red and blue with more general phrases.

"There are three special-purpose K-Js," says Maria. "The Chips for Thought is executing the K-J you think is best for your project."

Chips for Thought

1. Understand the three special purpose K-Js.
   Weakness or Problem Formulation K-J—This K-J is around a challenge. It is like answering the question, "What are the problems with…?" It doesn't get to brainstorm solutions. The focus is on the problems.
   Context or Image K-J—The focus of this K-J is on the word pictures that describes a situation or environment. This K-J may include images about the current context or future trends. The visual aspects are what matter.
   Requirements K-J—The focus of the requirements K-J is on functionality. For example, companies build a requirements K-J as part of their enterprise software selection process to capture the voice of the different stakeholders.
2. Which one is best for your project? Why?
3. Execute the K-J that you have selected.
4. What benefits do you perceive from your K-J exercise?

5S—"Marie Kondo"

The Netflix series Tidying Up with Marie Kondo shows different home situations where the expert Marie Kondo helps clients to clear out the clutter. 5S is like Marie Kondo, but in the workplace. It's a methodology to achieve a workplace that is clean, organized, and safe. The implementation of 5S brings the following benefits:

- Reduced cost—when companies can't find what they need because the workplace is not organized, they buy more.
- Higher quality—controls are better.
- Increased productivity—as all is where it should be, the team members can locate what they need faster.

- Greater employee satisfaction and retention—who doesn't like working in a nice, clean, and organized environment as opposed to dirty and cluttered? (Okay, maybe mice and other rodents prefer the latter.)

5S means the following:

Sort (Seiri)—Go through the items and keep what is needed and discard what is not.

Set in Order (Seiton)—After removing the non-needed items, arrange the remaining items for ease of use, based on the frequency of usage, space occupation, color of the objects, etc. It's important that the team can locate the items needed fast to perform.

Shine (Seiso)—Shine refers to the cleaning of the work area and regular maintenance on equipment and machinery. It's about housekeeping.

Standardize (Seiketsu)—"After the sorting, setting in order, and housekeeping, you don't want to come back to the original situation," Maria explains. "That's where standard operating procedures (SOPs) kick in."

Sustain (Shitzuke)—The translation of Shitzuke is discipline. Discipline is essential to preserve the accomplishments. It needs to become second nature to apply sort, set in order, shine, and standardize. "The book *Atomic Habits* by James Clear provides great practical ways to instill and keep the good habits," Maria states.

"And now, my fellow colleagues, a Chips for Thought with my purse," says Maria.

"With your purse?" asks Tim.

"Yes, you've heard well. You guys need to apply the 5S methodology and tell me about your discoveries," clarifies Maria.

"That's like putting your purse in order!" exclaims Jon. "I don't do that for my wife."

"I would hate it if my husband got in my purse because it is *my* purse," says Amy.

"Give this a chance…5S provides great benefits. If you transform a purse, wait to see what it does for your warehouses and workplaces," says Maria.

The team starts the Chips for Thought on 5S. It creates laughs, conversations, good times, and of course, all with good crunchy snacks!

Poka Yoke—Mistake Proofing

"PO-ka yo-KAY!" says Maria, like a warrior getting into a battle, making a fist, and pushing it forward.

Poka Yoke is mistake proofing. This process analysis tool implies a device, system, or method that makes it impossible for an error to occur or makes the error extremely easy to detect when it happens.

Poka Yoke comes in handy in the following situations:

- A process step that is prone to human errors or calls for mistakes. Maybe it yells for mistakes because it's challenging or boring.
- At a hand-off step in a process, when the output goes to a different worker, and to avoid a domino effect.
- When a minor error early in the process causes a hippopotamus error later.
- When the consequences of an error are big in money or safety.

"May I ask how Poka Yoke works?" asks Sarah.

"No question about it, Sarah," replies Maria, "let's see Poka Yoke in 5 steps."

Poka Yoke in 5 Steps

1. From the flowchart or process map, think about where and when human errors could occur. "A process map is a goldmine, don't you think, team?" Maria asks this rhetorical question.

2. For each potential error, go backwards to identify the root cause. "Think about the root-cause analysis tools like the fishbone diagram, 5 Whys, FTA, FMEA…" says Maria.
"How to forget the toddler approach and Maria's passion for trees?" jokes Ruben.
3. For each error, think how to make it impossible to happen. Some ways can be:
   - Elimination of the step or activity that is causing the error
   - Replacement of the step or activity for an error-proof one
   - Facilitation of making the correct action easier than the error
4. When you can't make it impossible for the error to happen, you need to look at inspection methods, setting functions, and regulatory functions. With inspection and setting and regulatory functions, you want to make detection as easy as a piece of your favorite cake.
   - Inspection methods—there are 3 main inspection types:
     - Successive inspection when the next worker performs it.
     - Self-inspection when the worker conducts it before moving forward.
     - Source inspection that takes place before the process step—on the setup.
   - Setting functions[22]—setting functions are methods to test for errors, including the contact or physical method, the motion-step or sequencing, the fixed-value or grouping, and information enhancement.
   - Regulatory functions—regulatory functions are signals that there was an error and include warning signals and control functions. Examples of warning signals are bells, lights, buzzers "like the ones you get at Olive Garden or other restaurants when you're waiting for your table," says Maria. Control functions stop the process until the team fixes the errors.
5. Choose the best mistake-proofing method; test; and implement it.

"What do you think about Poka Yoke? How about applying this process analysis tool for our on-time deliveries project? That's our next Chips for Thought. And you have the advantage that you have your process map, C&E, and FMEA matrices complete," says Maria.

Chips for Thought

1. Apply Poka Yoke to the on-time deliveries project.
2. What is helpful?
3. What has been challenging?

Kaizen

Kaizen means change (Kai) for the better (zen) or continuous improvement and can be part of lean six sigma. The 5 fundamental principles[23] that lead to continuous improvement are:

1. **Know your customer**—companies that identify the customer's needs and wants can deliver value for the customer.
2. **Let it flow**—the aim is to create value and eliminate the 3 Mus— Muda (wastes), Mura (defect/variation), and Muri (strain on staff and machinery).
3. **Go to Gemba**—Kaizen encourages being present where the actions that create value happen. "You don't want to be a spectator, you want to be there and run the movie or show. For those that love football, or sports in general, you want to be in the game. That's Gemba," says Maria.
4. **Empower people**—It's about organizing team efforts, setting goals, and providing a system and tools to reach them.
5. **Be transparent**—Data tells the story. Performance and improvements need to be visible and tangible.

The focus is on improving through the standardization of production processes. Kaizen uses Kaizen events called Kaizen Blitz. Kaizen Blitzes are cross-functional team meetings with the goal of completing specific tasks by the end of the meeting. The idea is implementing 80% during the meeting and the remaining 20% in the next 30 days.

Kaizen events are like HIIT (high-intensity interval training) workouts. HIIT is a training technique that has quick intense bursts of exercise (known as active periods) followed by recovery periods. A Kaizen event may last hours, days, or up to a week. They are quick, focused, and full of action, like quick punches—maybe double-punches—to get simple and sustainable improvements.

Kanban

Like Kaizen, Kanban can also be part of lean six sigma. There is some misunderstanding about Just-in-Time and Kanban. Just-in-time is the goal and implies visual management while Kanban is a tool to accomplish Just-in-time. There are other tools like Heijunka (production leveling), small lot production, and takt time in addition to Kanban.

Kanban is a tool within Just-in-Time that aims at reducing labor and inventory waste. It's a pull system. It means that companies regulate inventory based on actual customer orders. This contrasts with a push system where a forecast drives replenishment and production. Kanban has its origins in manufacturing in Toyota, but it has extended to services, with application for software development in agile environments.

Below are the key Kanban practices[24]:
- *Visualize workflow*—What does the workflow look like? Companies want to visualize the real workflow, not the ideal. This is the basis for the Kanban board.
- *Limit Work In Progress (WIP)*—This. Is. Critical. Like water, food, potato and tortilla chips are to life. "How about the tequila?" asks Ruben. "OK, we'll add the tequila for Ruben," replies Maria.
  By setting limits to the number of tasks, teams can keep the flow of tasks without the overflowing.
- *Manage the flow*—Bottlenecks happen. Effective teams address them by analyzing the underlying issues and not by adding WIP. Teams may divide the task in more steps or train team members in different steps to lower the dependence on specific team members.
- *Define each step*—A common ground facilitates the workflow. "Like with metrics, definitions are important for a common ground and alignment.

For example, Jon may have a different understanding than Mike of a task that has been completed," explains Maria.
- *Use feedback loops*—This includes daily huddles. These are standing up huddles to ensure that they are short and focused. They become popular with the scrum methodology.
- *Improve in collaboration*—Like many of the lean six sigma tools, Kanban takes place in a cross-functional team setting to boost participation and engagement of the team members.

"How does all this work in the trenches?" asks Mike.

"With Kanban boards and Kanban cards. Let's deal the cards!" says Maria.

Kanban Boards and Cards

The components are a Kanban board that contains Kanban cards. The Kanban board allows for the visualization of the overall workflow and tasks' progress. "You can also identify bottlenecks and potential issues. The magic word here is visualization. All about visuals," Maria emphasizes.

A Kanban board has tasks and columns as follows:
- backlog or to-do
- WIP by phase. For example, if the Kanban board contains the DMAIC phases, it shows a separate column for Define, Measure, Analyze, Improve, and Control.
- a "done" column.

"Is this like Trello when you move the sticky note from one column to another?" asks Tim.

"That's right, Tim; Trello and other project management tools like Asana also apply Kanban practices. What's the magic word?" asks Maria.

"Visualization," answers Sarah.

"You've said it Just-in-Time," adds Ruben.

Kanban Cards—Kan = Card; Ban = Signal

The Japanese word Kanban means signal card.

There are two types of cards: 1. Cards with tasks and 2. Inventory cards. Cards with tasks are straightforward. A team member should be able to complete the task in between two hours and two days. Inventory cards are those that are used to play the Just-in-Time game. With the inventory cards, teams know when they need to replenish inventory, like getting more cards from the deck.

"But what do you do with the cards? What are the mechanics?" asks Mike.

"This works like replenishing empty shelves in a store," Maria starts explaining.

When a team member needs a component, she goes to the designated area to get it. The racks there contain bins with the different components that a workstation uses. Each bin in this area has the requisition Kanban card fixed to it. This Kanban card is removable.

The requisition Kanban card contains, at minimum, the identification name, the SKU, the rack number, and the shelf on the rack where there are more bins with the component. The team member from the assembly line removes the card from the bin, hangs it in a hook from the rack, and takes the bin with the components to use to their workstation.

The requisition Kanban cards hanging from the racks are visible and show the need for replenishment. A team member going through the aisles can see the hanging cards and the empty bins on the racks. She takes the requisition cards and empty bins to the store or crib.

At the store, the team member replaces the empty bins with the corresponding full bins. Each bin in the store area contains a production card. Like the requisition card, the production card is also removable. For replenishment, the team member takes off the production card and includes the requisition card in the bin to use. A team member takes the now-full bins to the designated area for components.

To have the components manufactured each time replenishment takes place, the team member hangs the production cards in the manufacturing area. Manufacturing takes the hanging production cards, produces the components, and places them in the corresponding bins and shelves. The filled bins go to the store. If the team orders the components instead of manufacturing them, the same process applies, but with a supplier. Many Kanban systems are now components of an ERP that communicates in place of the cards.

"I can hear it, team! You are asking for a Chips for Thought!" says Maria.

Chips for Thought

Materials needed: index cards, any product, and 2 bins (one empty and one full). By using these materials, act out how replenishment works in a Kanban pull system.

Pugh Matrix
The Pugh Matrix facilitates decision-making based on defined decision criteria. "You can have different alternatives, or like in procurement, different suppliers to evaluate. The Pugh Matrix is helpful in situations where you need to select one or more improvement opportunities to proceed," explains Maria.

The steps to create the Pugh Matrix are the following:
- *Define evaluation criteria.* "With the procurement example of supplier selection, we need to have it before receiving the suppliers' proposals to avoid or minimize subjectivity," adds Maria. The voice of the customer or the voice of the business are inputs to the evaluation criteria.

- *Weigh your evaluation criteria by using a numeric scale or a relative percentage.* For some companies, quality or on-time deliveries carry a higher weight than price; for others pricing is king—or prince, at minimum—and of utmost importance. There are no wrongs or rights. What matters is that there must be alignment on the weights assigned to the different aspects and ranking methods in the evaluation criteria.
- *Determine your alternatives.* Following the example in procurement, the alternatives would be the proposals of potential suppliers and the new proposal from the incumbent (current supplier), if the incumbent participates in the Request for Proposal.
- *Define the baseline.* In most cases, this is the current situation. In the example, the baseline would be the existing capabilities of the incumbent. Companies define a baseline to understand the improvements.
- *Assess each alternative.* Based on the evaluation criteria, companies assess each alternative against the baseline. It can be better, the same, or worse. Companies also score each alternative. The alternative with the highest score is the best.
- *Explore better solutions.* Companies may go further to identify hybrid solutions with different aspects coming from more than a proposal.

"Procurement, what are your thoughts about applying the Pugh Matrix to supplier selection?" asks Maria.

"I think it would be helpful. Sometimes we get questions why we selected a particular supplier instead of others," indicates Sarah. "With the Pugh Matrix we can show the selection criteria and scoring."

"And dump that on their face," adds Mike. "That would also save us time in discussions that don't go anywhere."

"It's great to hear those comments, team!" says Maria. "Next we'll move on to some tools in the toolkit that can help with production."

Heijunka = Leveling

"I know that you guys at Alex's Snacks have co-manufacturers and not your own manufacturing plants. However, it's fundamental to assess the co-manufacturers' production capabilities and understand the different tools available," says Maria.

Heijunka is a lean method for reducing the unevenness in a production process and minimizing potential overburden. Like Kanban, Heijunka is a tool to accomplish Just-in-Time. Heijunka aims at optimizing capacity with short segments of standardized work to meet demand.

"Raise your hand if you are familiar with these situations that cause unevenness in production processes[25]," says Maria.
- Different tasks require different amounts of time—Everybody nods.
- Customer orders don't come in a uniform or ordered fashion—Ruben raises both of his hands up in the air, like a survivor on an island waving his arms high for a helicopter to discover him.
- Different employees have different capacities—"What a profound statement. Did it take long to figure this out?" asks Jon.
- "Agreed, Jon. This is kind of obvious," Maria replies.
- Lack of or limited resources to meet demand—"Here we go again," says Ruben, making the same movement with his arms.

"We have limited resources in every place. Just look at us. We work all the time," Tim adds.

"And believe me…I have been at Alex's Snacks for 9 years; it takes a long time and many approvals to go through to get hands to help out," says Mike.

"This is why Ernesto has asked me to help. Ernesto's vision is to have an elite supply chain and procurement team to drive profitable growth. We have frameworks, tools, processes to make it happen. Go team!" says Maria.

After a short break, the team resumes discussing Heijunka. Heijunka protects from overburden when demand spikes up. Overburden is a situation that the supply chain and procurement team at Alex's Snacks is way too familiar with.

To avoid overburden, production follows takt time[26]. Takt time is the product completion rate to meet customer demand. A longer or shorter takt time depends on customer demand, product assembly time required to meet demand, and net time available to work.

Heijunka has two ways to level up production:
1. *Leveling by volume*—"In a company like Alex's Snacks, starting new work when the customer order arrives doesn't work. Would you agree, Jon?" asks Maria.

   "Correct. That's why I don't get it why we're talking about Heishu or Heinu or whatever," Jon replies.

   "Because Heijunka has an absurdly good way to level up production in companies that receive a steady torrent of new customer orders," says Maria.

   Heijunka allows us to level production by the average volume of orders so we have a steady production flow. Instead of going with the flow with 10 orders on Monday, 1 on Tuesday, 5 on Wednesday and so on, it considers the weekly average to avoid the daily fluctuations.

2. *Leveling by type*—"Maria...we have many kinds of chips in different packaging sizes. Any suggestions for leveling production?" asks Mike.

   Maria knows that when companies are scaling, so is their portfolio of products, and this is no small potatoes at Alex's Snacks.

   Heijunka allows level production based on the average demand for each product in the portfolio and schedules the work around it by using a tool developed by Toyota, the Heijunka box.

   The Heijunka box[27] is a scheduling tool that facilitates visualization of the items to complete to meet the average customer demand. It is like a grid with the type of product in the portfolio and the amount to be produced each workday. Each item in the Heijunka box is a Kanban card that goes through production.

SMED—Single Minute Exchange of Dies

Related to production like Heijunka, Single Minute Exchange of Dies (SMED) is a lean six sigma tool to optimize production equipment. It applies to changeovers when the team goes from manufacturing one product to another one.

Maria remembers from her experience with the Walls at the Long Island New York Chocolates that changeovers were a big deal or, more in terms of

Alex's Snacks, a hot potato. The chocolate production team was attempting to optimize the changeover sequence because it is not the same going from light to dark than from dark to light. The multiplication rules where the order of the factors doesn't matter don't apply to changeover sequences.

SMED eliminates waste by reducing downtime and facilitates product changes with Just-in-Time manufacturing. Another important method to reduce downtime is Total Productive Maintenance (TPM).

Total Productive Maintenance (TPM)
Total Productive Maintenance (TPM) consists of the maintenance of production and quality systems that combine machines, people, equipment, and processes. Like the other lean six sigma methods, TPM aims at reducing waste, defects, failures, and losses, by using preventive and predictive maintenance techniques. Preventive maintenance uses historical data and best practices, while predictive maintenance takes current measurements and uses this data to callout potential issues with a proactive approach.

In productive maintenance, teams identify the critical pieces of equipment and assess their condition. This assessment is against well-defined scoring criteria that rank every piece in the organization. These scoring criteria set the ground for common definitions and priorities.

There is also an assessment of the maintenance skills to minimize downtime, identify any gaps, and potential training needs. This assessment also covers whether the company has enough in-house support.

"The implementation of total productive maintenance is rooted in…" Maria starts explaining. "Roots! Oops, I did it again, like Britney Spears's song says. I must love trees and gardening!" admits Maria.

The basis of total productive maintenance[28] is as follows:
- Safe, healthy environment
- Planned maintenance
- Equipment management
- Education and training
- Focused improvement

- Administrative and office TPM
- Autonomous maintenance
- Quality management

Control

"After going through the Define, Measure, Analyze, and Improve phases of DMAIC, we get to the Control phase, where we ensure that all the amazing improvements stay in place," says Maria.

The control chart in six sigma documents what to track, how, and who monitors to keep the improvements made over time and avoid going backwards or to the starting point.

The control phase is like the PDCA (plan, do, check, act) cycle—also known as the Deming cycle—a powerful tool in lean six sigma for continuous improvement.

PDCA (plan, do, check, act)

Teams follow the PDCA cycle in the following situations:
- At the start of an improvement project
- To control a cultural shift—"This is why we cover the PDCA tool in the control phase," clarifies Maria.
- Develop a new process or improve an existing one.
- Implement a new change.
- Plan data collection for root cause analysis

There are four stages:

Plan—This stage consists of understanding the problem and the impact on the process. In this stage, teams identify metrics to then measure the results.

Do—This is about action. It can be a test project to assess results.

Check—In this step, teams measure results.

Act—This step is act or adjust. If the test project was successful, teams can extend it to a larger scale. If not, they make the necessary adjustments and repeat the PDCA.

Standardized Work

Standardized work—also standard work—is a lean method that consists of finding the best current method of completing a process and making that method the standard. "It doesn't mean that the standard method will stay forever and ever. It adapts for continuous improvement," Maria says.

Standard work[29] has 3 key elements:
1. Takt time—Takt time is the production rate to meet customer demands. "How many potato and tortilla chips we need to produce per hour to meet our customer demand, correct?" asks Tim.
"I couldn't have defined it better," answers Maria.
2. Work sequence—This is the series of steps that work best to perform a process. Teams get at this order of steps after applying other lean six sigma tools for continuous improvement.
3. Inventory—"Unfortunately, I am too familiar with it," says Mike, "packaging, ingredients, WIP. You don't need to look further here at Alex's Snacks, many times, inventory is everywhere."

"And many times, we don't have the inventory we need," adds Tim.

Standard work is very helpful when there are many procedures that repeat and repeat, and companies need to ensure consistency. "This is a common problem in fast-growing CPG companies like Alex's Snacks," says Maria.

"Standard processes, SOPs, all that stuff is great if everyone follows; if they don't, it's a waste of time," says Jon.

"Excellent point, Jon!" Maria agrees. "We don't want to have waste, so we need to make sure that everyone follows the standard work."

To reap the benefits of eliminating waste, improving quality, and increasing efficiency, team members need to follow the standard work consistently, as Jon and Maria state. Developing instructional material plays an essential role in this, like forwarders in sports to make the score.

Effective instructional materials are simple and visual. Team members need to be able to access them and locate what they are looking for quickly. Patience and attention span are not in great supply these days. This explains the proliferation of brief TikTok videos and YouTube shorts (videos with a duration of less than a minute).

"Lean six sigma started in manufacturing, but it doesn't mean that instructional materials need to be boring. We can use our creativity and apply technology to have amazing videos and online courses," says Maria.

FMEA

FMEA is another tool companies use in the control phase, as it documents the current controls. Teams also conduct an FMEA in the Measure and Analysis phase of DMAIC.

"FMEA is in your six sigma template. Please feel free to revisit it with a good bunch of chips to make it a party," says Maria.

The control phase is the last phase in DMAIC. DMAIC addresses the current processes, while DMADV does so with the design process.

The 5-step DMADV Approach

DMADV[30] is part of Design for Six Sigma (DFSS) and applies to new process or product design and re-design. DMADV has five steps, sharing the first three steps with DMAIC:
1. D—*Define*—The define phase is to identify and set measurable goals based on the customer's expectations, organization, and stakeholders.
2. M—*Measure*—In this phase, the team defines the Critical to Quality (CTQ's) factors and quantifies them.
3. A—*Analyze*—The analyze phase includes building design alternatives and conceptual designs, assessing potential combinations of requirements, and selecting the best elements and designs.

4. *D—Design*—This includes a high-level and detailed design for the alternative that the team has selected in the Analyze phase. In this phase, the team assigns priority to the elements of the design and builds a high-level design to then build a more detailed design to prototype. The prototype helps to identify potential errors and adjust.
5. *V—Verify*—This is the final phase, where the team validates the design. They may need to create a pilot and try some production runs to ensure quality. This phase also includes a document with the transition or hand-off of the product to the process or product owner. Like in the control phase of DMAIC, in the verify phase, the team builds a control plan to sustain the change.

"With the SCOR framework we have mapped out the strategic goal tree to ensure alignment with the financial and supply chain and procurement metrics. We have also covered the lean six sigma methodology and the structured and data-driven approaches of DMAIC and DMADV. We have built a powerful toolkit. We are ready and hungry for profitable growth with increased revenues and cost reduction by addressing S&OP-Finance planning and execution processes," Maria summarizes.

Endnotes

1. Purdue University, *Six Sigma vs Lean Six Sigma: What's the Difference?*, 2021, https://www.purdue.edu/leansixsigmaonline/blog/six-sigma-vs-lean-six-sigma/.
2. USM Supply Chain, *Six Sigma Lean and Circular Economy with Guest Sneha Kumari*, 2022, https://usmsupplychain.com/six-sigma-lean-and-circular-economy-with-guest-sneha-kumari/.
3. USM Supply Chain, *Coronavirus and Supply Chain – Circular Economy and Supply Chain with Deborah Dull*, 2020, https://usmsupplychain.com/coronavirus-and-supply-chain-circular-economy-and-supply-chain-with-deborah-dull/.
4. USM Supply Chain, *Coronavirus and Supply Chain – Dana Drake-Cox, VP Global Supply Management Security Stanley Black & Decker*, 2020, https://usmsupplychain.com/coronavirus-and-supply-chain-dana-drake-cox-vp-global-supply-management-security-stanley-black-decker/.
5. Norman Vincent Peale, *Shoot for the moon. Even if you miss, you'll land among the stars*, Goodreads, https://www.goodreads.com/quotes/4324-shoot-for-the-moon-even-if-you-miss-you-ll-land.

6. Team Asana, *What is a Flowchart? Symbols and Types Explained*, Asana, 2023, https://asana.com/resources/what-is-a-flowchart.
7. Villanova University, *How Six Sigma Uses Failure Modes and Effects Analysis*, 2020, https://www.villanovau.com/resources/six-sigma/how-six-sigma-uses-fmea/.
8. UpKeep, *Fault Tree Analysis & Examples: What it is, How to do it, and Why it's Important*, https://www.upkeep.com/learning/fault-tree-analysis#how-to-do-a-fault-tree-analysis-in-6-steps.
9. TWI Ltd, FMEA VS FTA (*What are the Differences Between Them?*), https://www.twi-global.com/technical-knowledge/faqs/fmea-vs-fta.
10. Rebecca Bevans, *An Introduction to t Tests | Definitions, Formula and Examples*, Scribbr, 2020, https://www.scribbr.com/statistics/t-test/.
11. JMP Statistical Discovery, *The t-Test*, https://www.jmp.com/en_us/statistics-knowledge-portal/t-test.html.
12. Minitab, *What is a Test Statistic?*, https://support.minitab.com/en-us/minitab/21/help-and-how-to/statistics/basic-statistics/supporting-topics/basics/what-is-a-test-statistic/.
13. Minitab Blog Editor, *How to Correctly Interpret P Values*, Minitab, 2014, https://blog.minitab.com/en/adventures-in-statistics-2/how-to-correctly-interpret-p-values.
14. Lean Process, *Six Sigma Tools—Analyse Phase*. https://www.leanprocess.net/six-sigma-tools-analyse-phase/.
15. JMP Statistical Discovery, *The Chi-Square Test*, https://www.jmp.com/en_be/statistics-knowledge-portal/chi-square-test.html.
16. Keith M. Bower, *What is Design of Experiments (DOE)?*, American Society for Quality, https://asq.org/quality-resources/design-of-experiments.
17. Ken Feldman, *Taguchi Method*, iSixSigma, 2022, https://www.isixsigma.com/dictionary/taguchi-method/.
18. Minitab Blog Editor, *Regression Smackdown: Stepwise Versus Best Subsets!*, Minitab, 2012, https://blog.minitab.com/en/adventures-in-statistics-2/regression-smackdown-stepwise-versus-best-subsets.
19. American Society for Quality, *What is an Affinity Diagram?*, https://asq.org/quality-resources/affinity.
20. American Society for Quality, *What is an Affinity Diagram?*
21. David L. Hallowell, *Effective Use of Special Purpose KJ Language Processing*, iSixSigma, 2010, https://www.isixsigma.com/affinity-diagram-kj-analysis/effective-use-special-purpose-kj-language-processing/.
22. American Society for Quality, *What is Mistake Proofing?*, https://asq.org/quality-resources/mistake-proofing.
23. Masaaki Imai, *Definition of Kaizen*, Kaizen Institute, https://kaizen.com/what-is-kaizen/.
24. Ted Hessing, *Kanban*, Six Sigma Study Guide, https://sixsigmastudyguide.com/kanban/.

25 Deepthy Prabha, *12 Key Six Sigma Tools you Need to Fare Better in Your Career*, GreyCampus, 2021, https://www.greycampus.com/blog/quality-management/12-six-sigma-tools.

26 Kanbanize, *What is Takt Time and How to Define it?*, https://kanbanize.com/continuous-flow/takt-time.

27 Kanbanize, *What is Heijunka?*, https://kanbanize.com/continuous-flow/heijunka.

28 Deepthy Prabha, *12 Key Six Sigma Tools you Need to Fare Better in Your Career*.

29 Ted Hessing, *Standard Work*, Six Sigma Study Guide, https://sixsigmastudyguide.com/standard-work/.

30 Ken Feldman, *Define, Measure, Analyze, Design, Verify (DMADV)*, iSixSigma, 2018, https://www.isixsigma.com/dictionary/define-measure-analyze-design-verify-dmadv/.

CHAPTER 5:

S&OP (Sales & Operations Planning)/IBP (Integrated Business Planning); ERP; MRP

Key Takeaways

In this chapter, you will learn how to do the following:
- Know and Quantify the Impact of the S&OP/IBP process on the financial results—top and bottom line, cash flows, and asset utilization
- Identify opportunities in the S&OP process to deliver greater financial results and performance
- Define a sustainable and repeatable planning process to balance demand and supply—given the constraints—that turns into higher profits
- Perform Data analysis, Automate, and Leverage the functionalities of technology (including ERP, MRP, etc.), simulations/what-if scenarios
- Be prepared and equipped for S&OP cross-functional meetings to maximize their benefits

"Hold on, Maria. I see S&OP and IBP on the whiteboard. Which of the two? I think it would be helpful if you make up your mind," says Jon.

"You are raising a good point, Jon," says Maria, "there are different definitions around S&OP and IBP, and in general, around supply chain concepts, even the supply chain concept itself! There's an interesting article about that. Let me share the link."[1] Maria searches her computer and adds, "We start with the definitions for all of us to be on the same page and on the same paragraph too."

S&OP and IBP

S&OP stands for Sales and Operations Planning. Richard Ling and Walter Goddard, who were at the time with the management consulting firm Oliver Wight, coined the term in 1988 in their book *Orchestrating Success: Improve Control of the Business with Sales and Operations Planning*. The authors were working on production planning models and noticed that they needed to consider demand, production capacity, and supply. From there, Ling and Goddard developed a 5-step S&OP process with the following:
1. Gathering data
2. Sales or demand planning
3. Production and supply planning
4. A review meeting
5. S&OP approval at an executive meeting

As the focus is on the supply chain, a supply chain manager drives the process.

IBP stands for Integrated Business Planning. It extends beyond supply chain. Oliver Wight,[2] one of the pioneers in S&OP, indicates that IBP "represents the evolution of S&OP from the supply and demand balancing process developed in the early 1980s. Today it is a process that drives the alignment of all functions across an organization, models and creates readiness for alternate outcomes, drives deployment of strategy, and enhances collaboration across supply chains."

From Wight's words, a clear difference between S&OP and IBP is the scope. While S&OP aims to break silos within the supply chain to balance demand and supply, IBP aims to break silos on a much larger scale, at the enterprise-wide level. This implies a strong focus on strategy and finance. IBP goals and targets are financial and align with the company's financial budgets and long-term targets. A COO or VP Supply Chain or CFO drives the process.

As the scope is different, the execution of S&OP and IBP is also different. S&OP starts with the sales forecast in number of units. In contrast, IBP starts with the financial forecasts of the organization. The language or unit of measure is different. Sales talks about cases, physical units, while business and finance speak in monetary units.

"That's why you keep insisting on financial metrics and on measuring the impact on the financial statements," says Amy.

"Yes, Amy," agrees Maria, adding, "The starting point in the training was finance, not supply chain and procurement. The sequence of steps does matter." In Steve Jobs' words, you can't connect the dots looking forward; you can only connect them looking backward.

IBP has advantages over S&OP[3] in three main areas:
1. *Measures financial performance*—IBP focuses on profitability. S&OP's main objective is defining a viable production plan aligned with sales forecast. When it adds Finance as an extension, it may favor production volumes over profitability. When companies implement IBP in the right way, IBP shows performance in financial terms that aligns with the financial targets.
2. *Quantify the financial implications of supply and planning decisions*—S&OP plays a pivotal role in IBP as it relates to supply chain. IBP allows for quantifying the impact of supply chain alternatives on profitability and aids decision making. At the same time, these alternatives are workable from the supply chain perspective.
3. *Supports long-term strategy*—The horizon planning for S&OP processes is between 12 and 18 months. For IBP, the planning horizon aligns with long term strategic planning—extending over 18 months—and supports the short- and medium-term operational requirements.

"Pretty obvious. We're doing IBP at Alex's Snacks," says Maria.

"We run S&OP here and we have our challenges," Sarah interjects. "Do you know what results from S&OP—sorry IBP—other companies are getting?" asks Sarah.

"Love where your head is, Sarah," answers Maria. "You are looking for some benchmark data about IBP's benefits to compare."

IBP shows impressive results. McKinsey & Company's research[4] indicates that companies with an effective IBP process realize one or two additional percentage points in EBIT, service levels are five to 20 percentage points higher, freight costs and capital intensity are 10 to 15 percent lower—and customer delivery penalties and missed sales are 40 to 50 percent lower. McKinsey also states that IBP technology and process discipline can make planners 10 to 20 percent more productive.

In a similar fashion, Accenture's findings[5] show up to a 30% reduction in operational costs and a 10% reduction in working capital. Deloitte's research[6] on S&OP and IBP also exposes the benefits with a 6 percent growth in EBITDA, a 10% increase in on-time deliveries, and a 25% reduction in inventory, which improves asset utilization.

These averages are helpful for benchmarking. As not all companies start from the same situation, results vary. There are various maturity assessment tools to define the company's baseline. Among them, a well-known maturity assessment tool is the Gartner S&OP Maturity Model, which uses a scale from one through five. The more advanced levels correspond to IBP.

Most companies have a low maturity on Gartner's scale, between levels one and three. This means that the focus is on short-term (1 to 3 months) operational planning, and the objective is to balance supply and demand. Metrics suffer from the watermelon effect. They look green on the outside but red inside, with functions or departments competing against each other, instead of being aligned. In these low maturity levels, the focus is on internal functional efficiency and cost containment without looking at the customer, not even a sneak peek.

In the lower levels of Gartner's Maturity Model, the only participants come from the supply chain. No other functions join the party that, of course, has potato and tortilla chips at Alex's Snacks. Each function has its own planning cycle but then they are not connected, or the connection is spotty. For example, finance has its planning cycle, and sales has theirs too. The same applies to strategy and procurement, each with their own planning cycle.

Companies that get to balance supply and demand arrive at a functional solution for supply chain that may not be optimal from the financial viewpoint. For example, optimizing utilization may lead to higher inventory levels because of larger batch sizes. These challenges made S&OP evolve into IBP.

Companies in the higher levels in Gartner's Maturity Model have end-to-end visibility of their value chain that allows for better trade-off decisions across the business and for the benefit of the entire business, instead of focusing on supply chain only. "There is no question that finance and supply chain are key pillars in IBP," says Maria. "For real-life examples and stories about the intersection of finance and supply chain, check out this Adventures in Supply Chain podcast episode[7] with Guest John Ferraioli."

After sharing the podcast episode with the team, Maria hears slow clapping, with long intervals between claps. It comes from Jon, who says, "Spectacular presentation of life-changing IBP, Maria. What do we do from here?"

"Remember the SCOR framework that allowed us to build our strategic goal tree by mapping out the financial and supply chain metrics?" Maria answers.

"Yes," replies Jon.

"Okay, we went over a couple of limitations that the model has. Does anyone know the limitations?" asks Maria, looking at Amy, who is a fine note-taker.

"Checking my notes now, Maria," Amy replies.

In less than a minute, Amy says, "I've got them," and reads:
1. SCOR doesn't address sales and marketing, including demand generation, product development, and research and development.
2. SCOR doesn't include which process improvement activities companies need to implement.

"We are addressing both limitations," says Maria. "Team, this is the action plan."

1. Financial integration with S&OP to move towards IBP.
2. IBP dimensions including finance, demand, new products, supply, and executive.
3. Supply in depth with a holistic approach, considering the impact on the financial statements:
   a. What tools from our toolkit to use
   b. Potential data sources, where to get the data
   c. Successful approaches in other companies or best practices

"Quite an interesting journey ahead of us," says Jon. "I would even agree if it were before Covid. The pandemic showed that planning is crap. Sorry, excuse my French."

"Got you, Jon. You're not the only one thinking in that way. It looks like many supply chain practitioners have raised the same question because KPMG published an article in Supply Chain Quarterly titled 'Is it time to blow up S&OP?'"[8] Maria says. "You're welcome to read it, but the answer is not abandoning planning altogether, but making it better."

S&OP and IBP—honoring the definitions in this chapter—have changed amidst Covid-19 and post-Covid. For example, constraints on the material supply were uncommon before Covid. Nobody would have thought about supply chain disruptions with packaging materials. Prolonged lead times also became the norm because of shipping issues. There were times when some freight forwarders weren't taking shipments by ocean.

These challenges during Covid and then with the Russia and Ukraine war made planning teams adapt. For instance, procurement received the planning from production. Procurement then checked for availability of the required raw materials and packaging. If there were any constraints, procurement communicated them back to production, who made the necessary adjustments based on procurement's feedback and sent the revised plan to procurement.

Covid also made profound changes in demand planning, a critical component in S&OP and IBP. Consumer behavior—for example, hoarding toilet paper and other products during Covid—added complexity and confusion to the models that companies have been improving over time, based on historical data. It's not about getting rid of all planning; it's about adapting the planning.

"With this in mind, let's proceed with the financial integration in the path to IBP," suggests Maria.

Financial Integration with S&OP to Move Towards IBP

The Rise and Fall of S&OP[9] by Niels Van Hove highlights the importance of the financial integration. In his article, Van Hove indicates that after 30 years of development of S&OP, there has not been much progress in the most recent years. The author states that the reason is supply chain bias. For S&OP to survive, Van Hove suggests integration with other functions like strategy and finance. Both functions have been talking about rolling forecasting and budgeting since the 70s, much earlier than the birth of S&OP.

The evolved S&OP is IBP, where the financial integration is not an add-on. The financial integration is the foundation of IBP, or in Maria's terms, "It's the roots of IBP." Along these lines or branches, the strategic goal tree with the financial and supply chain metrics mapped out based on the SCOR DS model is the starting point.

The IBP concept requires a change in mindset for supply chain and procurement practitioners. Supply chain drives the traditional S&OP, but IBP is a company-wide effort, and P&L owners or budget holders ride this rodeo. Because of this change in ownership, the language is different.

At the executive meetings, participants talk about top and bottom lines, profitability, return on capital, and gaps to budget, as opposed to supply chain metrics like OTIF (on-time in-full), inventory, and forecast accuracy. The team makes decisions on trade-offs based on the P&L, on the financial impact. The what-if scenarios and alternatives are also in financial terms. "For example, it is not enough to say that we can have the oil on time for production if we expedite freight; we need to quantify and show the impact on the P&L," says Maria.

"As we have our strategic goal tree based on the SCOR framework, the toolkit from lean six sigma together with the structured approaches of DMAIC and DMADV, we are hungry for massive action. Our Chips for Thought turn into Chips in Action. Go Team!" Maria cheers.

Chips in Action

In pragmatic terms, Maria suggests the following steps:

- Together with the finance team, build and validate the strategic goal tree. That's like the Bible. As a quick recap, the strategic goal tree has metrics that align strategy with finance and with supply chain and procurement through the SCOR framework.
- Follow DMADV to design the IBP process together with finance.
- "I have that DMADV is for new processes or products and DMAIC is for improvements to existing ones," says Sarah.
  "I see Sarah's point. We are doing S&OP, Maria, so it's not a new process. It should be DMAIC," Mike says.
  "It looks like you haven't done the due diligence about our S&OP process," adds Jon.
  "I like this conversation and interest! I am aware of the current S&OP process at Alex's Snacks. As IBP requires starting with the financials from the P&L owner and using other language, we want a fresh start or clean sheet. IBP is a reset more than an incremental change. For these reasons, I'd recommend DMADV. Makes sense?" Maria asks.
  "Makes sense," confirms Mike and the rest of the team.
  "Perfect! Let me help you with the IBP processes, data, and people," says Maria.

IBP Dimensions, including Finance, Demand, New Products, Supply, and Executive

"In IBP, the first process steps focus on the P&L or budget and our strategic goal tree. What's next?" asks Amy. "I'd like to be more prepared to talk to the finance team."

"Demand planning," Maria answers. "In an overly simplified way, we have finance, demand, supply, finance again, and executive. In these processes within IBP, we have people, data, and systems, aiming at making the optimal trade-off decisions based on the financial impact."

Demand Planning

Michigan State University[10]—a top university in Supply Chain where Maria got her MBA—defines demand planning as a supply chain management process of forecasting, or predicting, the demand for products to ensure they can be delivered and satisfy customers. Demand planning addresses questions on the product volume and mix, the location of the customers or buyers of the product, the timing of the purchases, and the distribution channels.

Alex's Snacks has different products—flavors and sizes—in its portfolio, and multiple customers, with large retailers like Wal-Mart being the company's most important customer accounts. This situation is common in other CPG companies. Although those CPG companies have direct distribution (direct to the consumer) and distribute through marketplaces like Amazon, retail remains their most important channel of sales.

Demand planning includes three main aspects:
1. Statistical forecasting
2. Product portfolio management
3. Trade promotion management

Statistical Forecasting

Forecasting is predicting the future customer demand; this is the customer demand at a future point in time. It can be next week, next month, or next year. Short-term forecast—from one week to three months—relates to operational decisions. There are also medium-term forecasts—between three months and 12 months—and long-term forecasts—over a year. Companies create these longer-term forecasts for strategic purposes.

There are two main approaches to forecasting:
1. Qualitative approaches
2. Quantitative approaches

"Most companies apply a hybrid approach. How do you guys create your forecast?" asks Maria.

"I think we do it with a hybrid approach," answers Amy, looking at Sarah and Mike for confirmation. "We take the previous forecast and go over future changes with the inputs from sales and marketing. Getting to the plan takes lots of work and lots of number crunching."

"Agreed that this is not an easy task. It takes time to mature an S&OP process into an effective IBP," says Maria. "As you have a hybrid approach, let's cover both—qualitative and quantitative—to get to those 'aha' moments, team."

Qualitative approaches
Qualitative approaches include:

Grassroots approach
"Don't even mention it," says Maria. "Roots again! But this time there's a tweak. It's about grass and not trees," she jokes. "The grassroots approach is like Alex's Snacks is doing when talking to the marketing and sales teams to get their insights on future customer demand."

"If you hear those guys, Maria, they can always sell everything," says Ruben, "they are like Superman."

Maria chuckles and says, "We need to go deeper to understand and validate—or not—the inputs from sales and marketing, in particular if they tend to be over-optimistic. We need to keep in mind that all qualitative approaches are subjective assessments."

Market research

Another qualitative approach is market research. This approach applies to new product introductions (NPIs). It can consist of surveying customers, testing product performance, and having a focus group.

Executive judgment

Executive judgment is relevant for the longer-term forecasts that are strategic in nature. For example, to build a five-year forecast, companies leverage executive judgment.

Historical analogies

Like market research, this qualitative approach applies to new products. Considering how like items have performed, companies can estimate the demand for the NPIs, like a new flavor for potato chips.

Delphi Method

The Delphi method is a structured process that uses multiple rounds of questionnaires for experts to get a consensus opinion. After each round of questionnaires, the experts can see a summarized compilation of the last round. Every expert has the right to make changes based on the group response. "As you may note, this qualitative forecasting method puts together the individual and group expertise," says Maria.

"Any questions before moving on to quantitative approaches?" Maria asks.

After a short silence, Maria says, "Pretty straightforward, huh? With the quantitative approaches, we spice things up."

"I hope so," says Jon, rolling his eyes, "no 'aha' moments. Boring moments, I would say."

"Maria, you'll need to add Tabasco for Jon," says Ruben, relaxing the situation.

"I'm up for the challenge!" says Maria, and starts with the quantitative approaches.

Quantitative approaches

Quantitative approaches include two main groups: 1. Causal forecasting and 2. Time series forecasting.

"I'm ready to share a secret with you," says Maria, creating a suspenseful atmosphere like Stephen King in his horror novels. "You have the tools in your toolkit and already know a great part of causal forecasting."

"Where do we need to look? I can't find anything under causal forecasting. Can you, ladies?" asks Tim, gazing at Amy and Sarah for an answer.

"I don't find it either. Any luck, Amy?" asks Sarah, while Amy keeps thumbing through her notebook.

"Any trees or plants with that name?" asks Ruben.

"That's funny, Ruben. It's tricky to find causal forecasting because it has a different name in our toolkit. On top of that, we have various tools."

"Give us a hint," requests Tim.

"Or the answer, so we can be over with this," says Jon.

"Regression analysis and all the tools for root-cause analysis," says Maria.

"I knew it! It was about trees!" says Ruben.

Causal forecasting

As the name denotes, causal forecasting aims at identifying the causal drivers—a.k.a. predictors—of demand. The mathematical model—like regression analysis—allows us to predict demand when plugging in the values of the predictors.

Regression Analysis

There are different tools in the toolkit to identify these predictors (drivers or root causes), including the 5 Whys, the Fishbone diagram, the FMEA, and the FTA.

Causal forecasting includes regression analysis, a statistical method. Linear regression is a common technique of regression analysis. Linear regression implies one input variable and one output variable. Multiple linear regression implies more than an input variable (more than an x).

"Before Ruben takes a shot of tequila, we can revisit our example with on-time deliveries," Maria says, and reads, "What regression analysis does is give us an equation like this—please note, this is an example I've made up to explain:

$$y = 103 + 63x_1 + 13x_2$$

Where y = on-time customer deliveries
x1 = on-time packaging deliveries
x2 = on-time oil deliveries

You plug in the values for x1 and x2 and get the on-time customer deliveries."

A broader concept is multiple regression which includes linear and non-linear regressions with multiple independent variables (x's) or predictors. The start is with linear regression to keep it simple while possible.

Causal forecasting also includes Econometric forecasting. The Global Supply Chain Institute of the University of Tennessee indicates that Econometric forecasting[11] uses the interplay of data about demand with information on external elements that can move demand to create a demand plan.

Linear regression is the starting point for Econometrics, and ARIMA is one of the most widely used models to predict time series data like demand forecasting.

"Bear with me. We will see ARIMA in a moment," says Maria.

The University of Tennessee also advises that the implementation of Econometric forecasting is more complex, but these techniques provide more accurate results.

Monte Carlo Simulation
The Monte Carlo simulation is a mathematical technique used to estimate the possible outcomes of an uncertain event. It does this with an estimated range of values instead of a set of fixed input variables. It is a peek into the future.

Monte Carlo simulation works by modeling the probability of different outcomes in a process or system that cannot easily be predicted because of the intervention of random variables.

Maria uses the example from IBM[12] of calculating the probability of rolling two standard dice. To calculate this probability manually, a person would need to roll the dice 36,000 times, considering the 36 combinations of dice rolls and a sample size of 1000. The Monte Carlo simulation reduces the number of rolls by randomly sampling the possible outcomes, considering the 36 combinations and calculating the percentage of times that a person gets a particular number.

To run a Monte Carlo simulation, there are three steps:
1. Set up the predictive model—this means to identify the dependable variable or y, or sales, and the independent variables or x's. The x's drive the predictions.
2. Determine the probability distribution of the independent variables (x's). Teams can do so by using historical data or judgment or a combination of both to define a range of likely values and assign to each of the values probability weights.
3. Run the simulations with the random values of the independent variables. Repeat. Repeat. Repeat until there is a representative sample of the infinite number of possible combinations.

The greater the number of samples, the more accurate the sampling range and the better the estimation.

Clustering Forecasting Methods

Clusters are groups that share similarities based on attributes and patterns. Clustering is a segmentation of the data into manageable subgroups. "It is like when we categorize suppliers into strategic, business-critical, arm's length, or the ABC classification for items," explains Maria. "If they are in different categories, clusters, and groups, I'm going to work with them differently."

In demand forecasting, it's vital to identify the demand behavior of customers. From the historical data, teams can identify customer clusters or segments based on customers' behavior with the help of clustering algorithms, such as K-means, self-organizing maps, and fuzzy clustering.

Clustering improves the accuracy of the demand forecast because the predictions are for each segment that contains similar customers. When clustering methods are unable to identify a pattern for certain customers, they consider them outliers.

"Causal forecasting was short and sweet—or salty for you guys at Alex's Snacks. Shall we dive into time series forecasting?" Maria asks.

Time series forecasting

A time series analysis model implies using historical data to forecast the future. It is a look into the past to predict the future or extrapolate what occurred in the past to the future. The model looks for trends, cyclical fluctuations, seasonality, and behavioral patterns in the dataset.

These time series methods work well when future demand relates to historical demand, growth patterns, and seasonality. In their book Supply Chain Management: Strategy, Planning, and Operation,[13] Sunil Chopra and Peter Meindl indicate that the observed demand has two main types of components: systematic (S) and random (R). Systematic components measure the expected value of demand and include level, the current deseasonalized demand, trend, the rate of growth or decline in demand for the next period, and seasonality, the predictable seasonal fluctuations in demand.

The authors suggest that companies shouldn't focus on forecasting the random component because there's not much to do there; it is random. The focus should be on predicting level, trend, and seasonality instead.

"With this in mind, I'll show you some time series forecasting techniques. Please feel free to crunch them with potato and tortilla chips...You see...We learn. We eat. We have fun!" Maria says.

Naïve Forecasting Method
The naïve forecasting method indicates that the forecast for the next period equals the previous period. This means the next period's demand is the same as the last period's demand.

Simple Average
The simple average forecasting method stipulates that the forecast for the next period equals the arithmetic mean considering all periods. This means that the next period's demand is the average demand considering all periods. Every period has the same weight in the calculation.

Moving Average and Weighted Moving Average
The idea is that future demand is similar to the recent observed demand. The moving average forecasting method takes an average of a set of numbers in a given range while moving the range. This technique indicates that the forecast for the next period equals the arithmetic mean of the last set of periods. A set of periods can be any, like 3, 6, 12 months, etc. In moving average forecasting, all past observations have equal or uniform weight.

Weighted moving average is an average of moving averages. When we do this, weighted moving average replaces the oldest moving average with the most recent moving average, managing the data trend better than a simple moving average. A weighted moving average forecast assigns more weight to current data than older data.

"Makes total sense," says Tim, "the most recent data should have more impact on the forecast than data from 10 years ago."

"Back when I started it all was different. I wouldn't use a single number from that time," adds Mike.

"Maybe not for forecasting, but for the History channel," says Ruben, while the team gets ready for a quick break before the popular Exponential smoothing in time series.

Take 3: Exponential Smoothing; Double-Exponential Smoothing; Triple-Exponential Smoothing
"A smoothie? I'll have a peach smoothie, please." Ruben says.

"They have similarities, though," Maria says. "Exponential smoothing is a smooth, simple, and powerful forecast method within a time series. Like with smoothies, you have different options, three in this case: 1. Exponential smoothing, 2. Double exponential smoothing, and 3. Triple exponential smoothing."

"Good analogy, but what is it all about?" asks Jon. "I feel my head spinning with all these forecasting methods that I don't use. They are for demand planners and data scientists."

"If we understand the concepts, Jon, we are able to ask the right questions and achieve a better performance as a team, like the supply chain and procurement team," answers Maria and proceeds to explain exponential smoothing.

Like naïve forecasting and moving average, this method assumes that the future will be kind of like the past. The exponential smoothing model takes the level only from the demand history and then forecasts the future demand as its last estimation of the level.

"Level? I think I have heard that word before," says Mike.

"Yeah, here it is; the level is the current deseasonalized demand," says Tim, pointing at a slide showing the systematic components of the observed demand.

"The level is the average value of demand over time," clarifies Maria. The exponential smoothing model takes the last forecast that is based on the most recent demand observation and throws it into the future, like in the movie Back to the Future, in which the characters seemingly travel back and forth between the past and the future.

Sarah stares at her laptop's screen, rubbing her chin. After a while spent in silence, Sarah asks, "What's the difference between exponential smoothing and the naïve and moving average methods?"

"Yeah, I don't get it either. They seem to be the same thing," says Tim.

Exponential smoothing is better than naïve forecasting or moving average. A key reason is that this model assigns a tremendous weight to the most recent data. Put in a different way, the weight of each observation decreases exponentially over time. With moving average models, each observation has the same weight. Another difference is that outliers and noise have less impact compared to the naïve forecasting method.

"How about the other two smoothies, double and triple?" asks Ruben.

Exponential smoothing takes the level from historical demand, but it doesn't project any trends—the rate of growth or decline in demand for the next period. Double exponential smoothing addresses this limitation, bringing the level as well as the trend. It is exponential smoothing with trend.

Triple exponential smoothing recognizes seasonality, something that exponential smoothing fails to do. It is triple because it has three key systematic components: level, trend, and seasonality. The method is as follows:
- Decomposition of prior demand observations into base level, trend, and seasonal components
- Separate extrapolation of base level, trend, and seasonal components
- Re-aggregation of forecasts by summing base level, trend, and seasonal components

These three types of exponential smoothing, simple, double, and triple are also known as Holt-Winters models because of their authors.

"Cheers for exponential smoothing!" says Maria.

ARIMA

ARIMA stands for Autoregressive Integrated Moving Average. ARIMA is a type of model also known as the Box-Jenkins' method. Together with exponential smoothing, ARIMA is one of the most-followed approaches to forecast time series. The models have followers like in social media. Exponential smoothing and ARIMA would be like the LinkedIn's top voices.

While exponential smoothing methods are appropriate for non-stationary data, ARIMA models should be used on stationary data only. The ARIMA model predicts that the statistical properties of a stationarized series will be the same in the future as they have been in the past.

Stationary data means that the data have constant statistical properties including mean, variance, correlation, etc. This means that if with back to school, the consumption of potato chips is higher, the data is non-stationary.

There are some ways to make the data stationary. De-trending removes the underlying trend in the series. This applies to indexed data (data measured in currency linked to a price index or related to inflation) and to non-indexed data. Differencing can remove seasonal or cyclical patterns. Logging can linearize a series with an exponential trend when the series is not measured in a currency, like the demand in units. The limitation is that logging doesn't remove an eventual trend.

The intimidating name Autoregressive Integrated Moving Average indicates the three parts of the model:
- The Autoregressive (AR) part is a weighted sum of lagged values of the series.
- The Moving average (MA) part is a weighted sum of lagged forecasted errors of the series.
- The Integrated (I) part is a difference of the time series.

"You can see the ARIMA model as ARIMA (p,d,q) where:
p = the AR part
d = the I part and

q = the MA part," explains Maria.

ARIMA models are helpful for short term forecasting. The downside is the subjectivity involved in identifying the p and q parameters. How good the estimates of p and q depends on the skill and experience of the model developer.

"It looks like someone's just got a new toolbox with statistical models for forecasting," says Maria, making eye contact with each of the team members.

"Maybe this is the million-dollar or billion-dollar question, but I have to ask: what is the best forecasting method for us, for Alex's Snacks?" questions Mike.

Selecting the Right Forecasting Technique

When Maria is about to reply, Ernesto comes into the room. "Good morning. Getting here just in time to hear the revealing answer. What is it, Maria?" asks Ernesto, while pulling up a chair to sit closer to the team.

"With this rundown of forecasting techniques, you can feel like Handy Manny or Bob the Builder with a big toolbox full of these statistical models to use. I'm going to show a cheat sheet for you to help you make decisions about which model to use because there is not a single model that is the absolute best," Maria says.

"There isn't? That's too bad! I expected that answer, though," says Tim.

Companies want to consider the following factors in selecting a forecasting technique:

Purpose of the forecast and desired accuracy

They go hand in hand. "You need to define the level of accuracy that you can live with depending on how you're going to use the forecast," Maria explains.

For example, a forecast for budgeting purposes should be more accurate than a forecast to decide if to enter a new business. In the latter option, a simple estimate of the market size may suffice for the purpose.

Likewise, required accuracy is different for a forecast to evaluate performance than for a forecast for planning use. "If management wants to know the effects of a special promotion in sales growth, we want the accuracy level to be high because of the impact on the top line of the P&L," Maria says, looking at Ernesto. She adds, "The level of accuracy we want depends on the impact on the financial statements."

The impact on the financial statements defines what forecasting technique to use, not the forecast accuracy.

Regarding forecast accuracy, companies need to keep in mind the following five points:
1. **Forecasts are always wrong.** Nobody can argue that. For that reason, forecasts need to include the expected value and a measure of forecast error. There's a plethora of ways to measure forecast errors, but most of the forecast accuracy metrics are variations of the following three:

- *Forecast Bias*—Forecast bias is the difference between forecast and sales. If companies get a positive result, that means that they are overestimating sales. "Like the marketing folks," interjects Ruben. When companies get a negative result, they are underestimating sales.

The calculations of forecast bias are as follows:

$$\text{Forecast bias} = \Sigma \text{ (Forecast} - \text{Sales)}$$

$$\text{Forecast bias \%} = \Sigma \text{ Forecast/Sales}$$

It's helpful to define if demand is systematically over or under-forecasting. This may not cause too big of a deal in the replenishment of a particular store, but this systematic error can cause the distribution center to be overstocked, considering that the distribution center covers multiple stores or doors.

When there are aggregations of several products or of periods of time, companies can give themselves a pat on the back, considering the bias of the overall forecast. But, at the detailed forecast level, bias can be ugly.

- MAD—It stands for Mean Absolute Deviation. MAD shows how far we are from actuals. It is the absolute value of the difference between forecast and actual sales. For example, considering a period only, if the forecast value of sales for the cheese curls that Ernesto worries so much about is $10M, and actual sales are $8M, MAD is $2M. If the actual sales are $12M instead, MAD is still $2M because it is the absolute value.

The calculation of MAD is as follows:

$$MAD = (1/n) \Sigma |Forecast - Sales|$$

MAD doesn't help with comparisons because it uses units (physical or monetary) and not percentages. For instance, an average error of 13,000 units is huge like a T-Rex, a dinosaur, not a statistical distribution—if the product sales are 15,000 units. This average error, instead, would be little as a ladybug if the product sales were 15,000,000 units.

- MAPE—It stands for Mean Absolute Percentage Error. MAPE addresses the limitation of comparison that MAD has, as it shows the average error in percentage points. MAPE is one of the most widely used forecasting metrics in demand planning.

The calculation of MAPE is as follows:

$$MAPE = \frac{(1/n) \Sigma |Forecast - Sales|}{Sales}$$

As a word of warning, MAPE calculations can show large error percentages with several slow sellers in the dataset. This happens because MAPE assigns the same weight to all the products and the relative errors with slow sellers seem much larger than in absolute values. For example, if the forecast for a slow seller is ten units and the actual sales are a single unit,

MAD is nine while MAPE equals 900%! This issue worsens when there are no sales, preventing companies from calculating MAPE.

The calculations of forecast bias, MAD, and MAPE are simple arithmetic calculations, no rocket science. For forecast bias, the target level is 1 or 100%; and for MAD and MAPE, as both measure forecast error, the target level is 0 or 0%.

2. **Aggregate forecasts are more accurate than disaggregate forecasts**. This applies to aggregated forecasts over products (product group level forecast) and over time; for example, a weekly forecast is more accurate than a daily forecast. This is because larger volumes hinder the impact of random variation.

Reassuming the calculations of forecast bias, MAD, and MAPE, things can get muddy with the aggregated level forecasts because there is more than one way to perform the MAPE calculations: 1. Calculate the MAPE of each product and do the average for the group, or 2. Apply the formula to the totals of the group (total forecast and total sales), not considering MAPE at the product level.

"Do we get the same result if we do either 1 or 2?" Maria asks the team. "I hear yes, no, silence, I don't know, I don't care. The answer is that we get different results, and we get a better MAPE if we follow option 2."

"Does it mean that we need to calculate MAPE in that way?" asks Amy.

"We can't say yay or nay because we choose option 1 or 2 based on what we expect to get out of the forecast."

If the forecast is to make decisions at the aggregated level, like planning the resources at the distribution center or production for the Superbowl or Cinco de Mayo seasons, companies consider the aggregated level. If the decisions to make are more related to a specific store, using the detailed forecast would be more appropriate. The same logic applies when aggregating over periods of time.

With aggregated forecasts, teams can choose between arithmetic or weighted average forecast. With regular MAPE, teams calculate the arithmetic mean, with each value having the same weight. Volume-weighted MAPE is a variant of MAPE that considers the weight of each product over sales. In so doing, the high sellers have priority over slow sellers.

3. **Long-term forecasts are less accurate than short-term forecasts**. "This is like with the weather forecast. It's fine for today or tomorrow. If we want to know how the weather will be in 10 days, we don't trust the forecast so much," explains Maria, smiling, "and as forecasts are always wrong, I keep an umbrella in my purse."

4. **Forecasts are more accurate for high sales volume.** This means that it is easier to build forecasting models for Alex's Snacks than for mom-and-pop stores. High volumes make it easier. "A brutal example that illustrates this point is that it is easier to forecast Doritos than a brand little known in the market."

5. **Bullwhip effect.**

"It's time for…" Maria starts saying when Mike anticipates "Chips for Thought" or the new "Chips in Action" and says, "I knew that was coming."

"I'm sorry. Right now, I need to attend another meeting," says Ruben.

"What meeting?" asks Ernesto with a surprised look while looking at his calendar.

"I'm teasing Maria, Ernesto," says Ruben.

"Actually, guys, it's a game," says Maria. "It's the beer game."

"Oh, I have just received a meeting cancellation notice. I'm staying," jokes Ruben.

"It's not with beer or any other substitutes," clarifies Maria, while pulling out some cards and markers from her bag—the same bag that the team organized following 5S. "Ready for the action?" she asks.

Jay Forrester, a professor at MIT, invented the beer game to show the bullwhip effect. The bullwhip effect refers to a supply chain phenomenon that small fluctuations at the retail level can cause larger and larger fluctuations in demand at the wholesale, distributor, manufacturer, and raw material supplier levels. The farther from the end consumer a company is, the greater the forecast error. "This is what we are experiencing right now full-blown, with issues with excess inventory after Covid," Maria says.

Cost

Once companies decide on the accuracy or inaccuracy they can live with, they need to make decisions on cost. It's a trade-off between cost and accuracy in selecting the forecasting technique. "Related to the excess inventory situation that Alex's Snacks is facing, like other companies, the dilemma is between the cost of more sophisticated techniques or the cost of having inventory in excess," says Maria.

"It's not only the cost of the forecasting solution that matters but also defining where the company is in its transformation journey. I explained this to the Walls, Ernesto."

Their company—Long Island New York Chocolates—wasn't using many statistics for demand forecasting. It would have been a stretch to go from where they were to machine learning (AI solution) in a leap. This podcast episode talks about the phases or roadmap to digital transformation[14] in supply chain.

Availability of historical data

Data gathering is the first step in the S&OP and IBP process (the evolved version of S&OP). Countless books and materials over the internet show that S&OP and IBP kick in with the collection and preparation of the data. This is not the order that Maria follows with her explanations.

Maria goes over the forecasting techniques first before diving into the data. The reasons are two-fold:
1. Without providing the toolset with the forecasting methods, teams wouldn't have a way to know what data they need to collect, provided that there is historical data.
2. Teams use data in all the IBP processes, not only for the demand planning stage.

Regarding historical data, it is helpful that teams obtain and prepare data about historical sales—in physical and monetary units, by product, by product groups, by customer, by customer groups, by location, and by channel. This segmentation of the data provides insights into trends and patterns. "Remember clusters?" asks Maria. "Keep them in mind."

It's important to note that sales don't necessarily equal demand. A company may have stockouts and backorders. This is unsatisfied demand, not reflected in the historical sales that need to be considered to forecast future demand.

There are two main situations that could happen when teams go fishing or hunting for the data to create a forecast:
1. There is an ocean of data with waves as high as those in Hawaii or Australia, whirlpools, and plenty of fish of different sizes and shapes.
"From trees and gardening, we are getting into surfing?" asks Ruben.
"See? I'm being creative, team. We have added oceans to the trees! Trees are coming back, though," answers Maria.
2. There's a desert of data, maybe with some tiny lakes and little fish to catch.

When companies are inundated with data in the ocean situation, they can apply the Pareto rule by getting the 20% of the data, which creates 80% of the impact on the financial statements.

Knowing what NOT to focus on is equally or more important than knowing what to focus on. James Clear, the #1 New York bestseller author of Atomic Habits, has a blog post about Warren Buffet's 2 list strategy.[15] Warren

Buffet's pilot, Mike Flint, shares the 3-step productivity strategy that Buffet uses to set up priorities:

Step 1—Write down the top 25 things that you want to accomplish this week.

Step 2—Circle your top 5.

Step 3—In step 3, there are 2 lists, one with the 5 circles and the other one with the remaining 20 goals. Buffet indicates to Flint that his list with the 20 goals becomes his Avoid-At-All-Cost list until he finishes his top 5 goals.

This productivity strategy applies to working with data for demand forecasting.

Selecting that 20% of data is not trivial. Data sources are the ERP (Enterprise Resource Planning) system, CRM (Customer Relationship Manager), data from spreadsheets, and databases. Lora Cecere takes an interesting angle in her blog Supply Chain Shaman[16] about considering channel data for the demand planning process.

Cecere highlights that there are still empty shelves in stores due to planning processes that are fundamentally broken. Cecere argues that new technologies like NoSQL—Graph and Cognitive Ontological learning—models can manage several data sources and produce a forecast in role-based views for finance, promotion planning, manufacturing planning, transportation planning, and procurement. These technologies model channel data with order and shipment data.

In contrast, traditional planning allows for one input to generate one output for the forecast. Traditional planning doesn't use or, better said, can't use channel data. These systems rely on order or shipment data that can't synchronize with channel data because these couple of data sets differ in planning horizon, granularity, and unit of measure. It is as if order or shipment data has communication challenges with channel data, and NoSQL would be the therapist to help.

Cecere concludes her article by stating that the use of channel data improves demand accuracy by 40-60% and reduces the bullwhip effect and demand latency. The article defines demand latency as the time from shelf purchase to translate the sale through backroom replenishment to warehouse replenishment to a retail order to a manufacturer. "I would love to see the financial impact of the 40-60% improvement in demand accuracy that Lora states," says Maria.

"So do we need NoSQL for more reliable forecasts?" asks Ernesto.

"It's a trade-off decision," answers Maria. "Let's talk more about the applications out there."

When instead of an ocean, there is a desert of data, companies need to consider other models than time series ones like moving average, exponential smoothing, and Box-Jenkins that require two full years of history, at minimum. Some causal models require several years' history. In these situations, companies use qualitative forecasting techniques.

Availability of forecasting software

There's hype around Artificial Intelligence (AI) and the power that robots can bring, those arms and legs moving around, dusting off and fixing everything they see in their way. In more pragmatic terms, AI is machine learning.

Traveling back in history to year 1959, Arthur Samuel, an IBM employee and pioneer in computer gaming and AI, coined the term. Later, in the late nineties, Tom M. Mitchell provided a formal definition of what machine learning entails: "Machine learning is the study of computer algorithms that allow computer programs to automatically improve through experience."

Mitchell defines these algorithms as "a computer program is said to learn from experience E to respect to some class of tasks T and performance measure P if its performance at tasks in T, as measured by P, improves with experience E."

Machine learning works like training a dog. Maria's husband, Jim, is a dog lover. When the kids were little, he convinced Maria to bring home Max, a golden retriever puppy, because the kids "needed" a pet to play with and to become more responsible, as they would need to take care of the dog.

Fast forwarding a few years, puppy Max became an 85-pound dog, the kids don't take care of him, and the person who enjoys Max the most is Maria's husband. Jim considers Max a smart dog because he does tricks like barking when requested, turning around, standing up, and rolling over.

Max has learned to perform all these tricks thanks to Jim's training. The training consists of verbal commands and signals. For example, Jim says "roll over" while moving around his right index finger. In machine learning, Max would be the algorithms, Jim's training would be the experience E, the verbal commands and signals would be the training data set T, and the tricks would be P, the predictive model in demand planning or the result.

"First things first. We want to start with the definitions of some terms to avoid confusion," Maria indicates. "Do me a favor and google machine learning. On the sites listed, you can see IBM.[17] There we have useful definitions and distinctions."

Machine learning, deep learning, and neural networks are subdivisions of AI. IBM goes further to explain that neural networks is a sub-field of machine learning, and that deep learning is a subfield of neural networks. The difference is on how the machine (or Max) learns.

The School of Information at UC Berkeley[18] explains how algorithms learn and distinguish three main parts:
1. A decision process—Algorithms make a classification or prediction. This is the output. To do so, they use input data that can be labeled or unlabeled. If the data is unlabeled, it's the algorithm that estimates a pattern. Deep machine learning can use either labeled or unlabeled data sets. Labeled data is also known as supervised data because it requires preparation and more human intervention. Unlabeled data—also known as unsupervised data—is raw, like text and images.

Traditional machine learning can use supervised data. It is us—the humans—that determine the set of features to understand the different inputs for the machine to learn.

The deep part in deep machine learning refers to the number of layers in a neural network. If the neural network has more than three layers, it's a deep learning algorithm. With less than three layers, it is a basic neural network, belonging to the classical machine learning.

2. An error function—As the algorithm predicts, the error function can make a comparison with actuals and assess accuracy of the model.

3. A model optimization process—The model aims at identifying whether there is a better fit to the data points in the training set, based on the actual observations, and adjusts the weights accordingly. The algorithm repeats the evaluation and optimization processes until it reaches a certain accuracy threshold.

"I'm pulling back the curtains, team," says Maria, like making an earth-shaking announcement. "You have machine learning in your toolkit. In fact, you have seen it in action."

The team members look at each other cluelessly, even Ernesto.

"How about regression analysis?" asks Maria.

"Is that machine learning?" asks Mike, emphasizing the word "that."

"Yes, it is," confirms Maria. "Linear regression and logistics regression[19] are both part of supervised learning models. These supervised models use labeled data. Linear regression is to predict, as we've seen, and logistics regression is more related to classification."

Other models included under supervised learning are neural networks, naïve bayes, vector machine (SVM), decision trees, and random forest.

"Hey Ruben, trees again! And we are going big with forests that combine several decision trees," Maria says, kidding Ruben.

Another method is unsupervised learning models that can work with unlabeled data. They include neural networks, k-means clustering, and probabilistic clustering methods. "Clusters! We have that as a forecasting method for demand planning. Am I right?" asks Sarah.

"Outstanding, Sarah! Excited to see that we are getting the 'aha' moments," says Maria.

There is also a mix or blend of supervised and unsupervised learning. This machine learning method is known as semi-supervised learning. It uses both a small, labeled data set, and a large unlabeled data set. This method becomes handy when there is not enough labeled data for a supervised learning algorithm.

"These are the main three machine learning methods. Reinforcement machine learning is similar to machine learning, with the difference that the algorithm uses trial and error instead of sample data. The IBM site states that the company won the 'Jeopardy!' challenge in 2011 by using reinforcement learning."

"Maria," says Jon, pausing for a few seconds to organize his thoughts and words, "all this bluff about machine learning and nothing about technology. Are you kidding me?"

"Too long of a context?" Maria asks, "I think it is important to understand concepts, logic, processes, and applications before talking about technology, but we are there now."

The infamous spreadsheets in Excel are the most widely used tool for planning and statistical analysis. Excel has its lovers and haters. In their marketing materials, technology companies mention eliminating all the spreadsheets around—like fumigating against a plague of nasty bugs—as a key motivator to go through a transformation.

The truth is, companies can't get rid of all the spreadsheets; some survive, like Gloria Gaynor says so vocally in her song, "I Will Survive." This is not horrible. In some cases, there is not a better option, and these Excel spreadsheets cohabit with sophisticated ERPs systems and APS (advanced planning systems), including solutions from SAP, Oracle, JDA, Infor, and Epicor.

Excel has an add-in named "Data Analysis Toolpak" that allows for statistical analysis. For example, companies can run triple exponential smoothing (the time series model that recognizes level, trend and seasonality) even if they don't know the calculations behind it. In the top ribbon, there's an option, "Data," that displays "Forecast." By selecting the data and clicking on that button, Excel automatically generates a forecast by using triple exponential smoothing.

Excel is super powerful and allows for advanced statistical analysis performed in costly software packages like Minitab. In addition, automation with VBA empowers teams by providing more accurate data fast, as it minimizes or eliminates manual data inputs and manipulations. For data visualization, teams can create dashboards and reports in Power BI.

"Excel is amazing, and many of its treasures remain unhidden for many companies. There's untapped potential in Excel," says Maria, who is on the love-Excel side of the equation.

"Can we do regression analysis now that we know that is machine learning in Excel?" asks Amy.

"Yep," answers Maria and adds, "I have seen several large companies using Excel for data analysis and planning and some companies using Microsoft Access (SQL language)."

"I have heard about python, pandas, and anaconda for machine learning. What are they?" asks Tim.

"True. We have a complete zoo for machine learning," jokes Maria. Python is a commonly used programming language that companies use for data analytics and machine learning. R is another statistical programming language—popular too but less than Python—that applies to machine learning. Anaconda is a distribution of packages—including python and R—built for data science, and it includes over 100 Python packages by default. Pandas is an open-source Python package, amongst others, for data analysis or data science and for machine learning tasks.

"Good explanation about the zoo for machine learning!" Ernesto encourages Maria.

"Thank you, Ernesto, for your words! I'd also like to mention that Sap, Oracle, and JDA offer solutions comprehensive of the IBP process in its entirety. Demand planning modules leverage the use of algorithms for predictive analytics."

Time needed to gather and analyze data and prepare a forecast

Another factor to consider in choosing a forecasting technique is the time it takes to collect and analyze the data.

With qualitative forecast methods such as the Delphi method and market research, in general, building a forecast takes between two and three months. In contrast, the panel consensus—another qualitative method—takes about two weeks.

Within quantitative methods, time series analysis, including moving average and exponential smoothing, takes less than a day. Because of this benefit and its simplicity, the time series forecasting method is a favorite one for many companies. The Institute of Business Forecasting (IBF) indicates that almost 60% of companies use time series with their order and shipment data.

Causal methods, also within quantitative methods, can take several weeks or months. For instance, teams build an Econometric model in about two months or longer. With regression analysis, the time it takes to prepare the forecast depends on the ability to identify relationships.

Considering quantitative and qualitative methods, a forecast can take as little as less than a day to three months or more.

Forecast horizon

Lora Cecere, in the same blog post about channel data, defines different time horizons. Cecere suggests aligning the thinking with supplying lead times. A model that extends beyond lead time is tactical, a model within lead time is operational, and a model within the order duration is an executional signal, as Cecere puts it.

The tactical and operational forecasts are not synchronized. Retailers cover from three to five months with their planning, while CPG companies extend their forecast to between 12 and 18 months. Granularity may also be different, as retailers do their planning in weeks and CPG companies use months.

Depending on the time horizon, some forecast methods are more appropriate than others. Qualitative forecasts are better than quantitative for longer-term forecasting. For shorter-term forecasts, ARIMA or Box-Jenkins would be a good option, for instance. Causal models like regression analysis and Econometric models are appropriate for short-term as well as long-term forecasts.

Life-cycle stage of forecast for a particular product

If the forecast aims at predicting demand for a particular product, it needs to consider the life cycle stage.

The stages in the product life cycle[20] are as follows:
1. *Introduction*—In this stage, companies launch the product into the marketplace and generate demand for it. There are low sales that increase slowly. Costs are high.
2. *Growth*—In the growth stage, customers have accepted the product and there is an increase in market share. Competition builds up too, while costs go down significantly.
   "This is the case with our Original and Nacho Cheese chips," says Mike. "I would say that this is the stage where most of our potato and tortilla chips are," Ernesto adds.
   During the growth stage,[21] the company's challenge of "getting consumers to try the product" becomes "making consumers prefer its brands."

3. *Maturity*—In this stage, the product is established in the market. Companies enjoy peak sales of the product. Costs are low. Products can be in the maturity stage for months or years or decades or centuries.
   "The Coke that you are drinking, Jon, is an example of a mature product for over a hundred years," says Maria.
4. *Decline*—In the decline stage, revenues go down because of market saturation, high competition, and changing customer needs. Cost is higher and profitability goes down.

Traditional quantitative forecasting techniques use the straight-line or growth rate to apply to quantities or volume sold, price assumptions, or both. An alternative approach is an analysis and forecasting of new product growth rate based on the S-curve (named because of its shape) that considers the phases of the product. Life-cycle models can predict the typical sales pattern, how sales evolve when the product gets to a certain age, and when competition gets intense.

Considering the product life cycle and acceptance by the different groups—innovators, early adopters, early majority, late majority, and laggards—allow companies to generate more realistic forecasts at the product level.

Product Portfolio Management; SKU Rationalization; Product Portfolio Optimization

Related to product life cycle, product portfolio optimization is more important than ever to cut costs and to increase revenues and profits. Product portfolio strategy and SKU rationalization is not only what to offer, but also when to offer the products, what customers, what channels, and at what prices. A multi-dimensional analysis considering all variables can allow for true portfolio optimization.

Key Areas and Challenges
Product portfolio optimization[22] consists of two main aspects:
Rationalization of products and product lines—Product rationalization involves decisions on which products and product lines to keep and which ones to kill.

Effective demand management—Demand management implies applying the 4Ps—product, promotion, place, and price—that Jerome McCarthy created back in the 1960s.

"How do you guys start with portfolio optimization at Alex's Snacks?" asks Maria.

"Sales and the brands identify the low sellers," says Amy.

"I think we can do better than that by taking a different approach. Please don't take me wrong; you guys are ahead of the pack by addressing portfolio optimization as part of the S&OP cadence, but we want to start with low performers from the profitability perspective, instead of low sellers," explains Maria.

In its research published in 2020, McKinsey[23] shares an example of a consumer product company that was focusing on increasing revenue, paying no attention to profitability. In a three-year period, the number of SKUs increased by more than 50%, the sales per SKU reduced by over 30%, and margins went down to about 10%. A rationalization program that included portfolio optimization, product design, and commercial-network alignment reduced the business portfolio by 25% while improving gross profit by 3%. McKinsey's example is one of the many companies that have under-optimized product portfolios with net-negative effects on costs, market share, cash flows, and profit margins.

Effective product portfolio optimization requires a systematic and rigorous approach. Mark Covas,[24] former planning director at Coca-Cola and J&J, describes in detail ten rules for product portfolio optimization. Covas suggests that companies should divest low margin brands, caring less about their size.

Other companies like General Mills and Procter & Gamble share Covas's thoughts and are taking the suggested approach. By doing so, companies can allocate their marketing dollars in a more productive way—from the low performing products or Pac-Man profit eaters in Ernesto's terms, to the high performing ones.

"Tired of PepsiCo, Maria? Because we are now having examples with Coca-Cola," says Ruben.

"I like variety, as most consumers do," replies Maria.

"Tell me about it," says Ernesto. "Because of changes in the packaging sizes here and there to capture market opportunities, and because of the creation of new flavors, we end up with many SKUs and a portfolio that keeps growing."

"Unfortunately, it is a common situation in fast-growing CPGs. The team and I will fix it," Maria says.

"Trust me, Ernesto. Maria is an expert in pruning," jokes Ruben.

There are two main hurdles on the rocky path to portfolio optimization:

Politics
The different functions or departments in a company can suggest conflicting product mixes because of misaligned goals. For example, the sales folks in the pursuit of revenues and happiness, because of performance bonuses tied to targets, may give preference to products with high turnover or velocity.

The same applies to other functional teams. Marketing looks at products to extend the company's footprint, while supply chain and procurement sneak up everywhere to slam cost, like birds of prey. Finance watches for costs, revenues, profits, and cash flows like a bird mom watches her chicks, but a great short-term decision may not be the best option considering long term goals and constraints.

"What can we use to address the political hurdle and align the functional teams?" asks Maria.

"I think I know the answer," says Ruben with enthusiasm.

"I think I do too," confirms Tim.

"Me too," says Mike, while Jon, Sarah, and Amy are nodding.

"The only one that doesn't know is me? Did I miss the memo?" says Ernesto.

"Team, can you say the answer for Ernesto at the count of three?" asks Maria. "One, two, and…three!"

"Strategic goal tree," says the team beautifully, like the sound of a highly acclaimed orchestra playing a Beethoven or Mozart masterpiece.

The strategic goal tree with the strategy, financial, and supply chain metrics mapped out provides solid ground for adding the metrics of the other functions or departments to achieve alignment and march together at a single rhythm.

Emotions

Another hurdle in the quest for portfolio optimization is emotions. "You don't know how many times I have heard: I can't believe it! You are going to discontinue product dot, dot, dot…I love it!" says Maria and adds, "Or the packaging engineer arguing for a new packaging with all the bells and whistles and a whole band, totally ignoring customer demand."
Defining and agreeing on decision-making criteria in advance helps to minimize subjectivity and keep emotions cool. The Pugh matrix, within the lean six sigma toolkit, is a helpful tool to define such criteria. In the evaluation criteria for portfolio optimization, companies want to include aspects about performance or profitability and shared components of finished goods.

Performance or Profitability (Not Sales)

Portfolio management is challenging, as it is not merely reducing complexity. Complexity can be good and can be bad, like good and bad cholesterol. McKinsey[25] makes this distinction about complexity. Good complexity creates a positive impact on profitability because the customer is willing to pay for the variance. Bad complexity hurts profitability with a negative net margin or an insignificant increase in profitability. An example of bad complexity is keeping a large legacy portfolio in case a single major customer wakes up one day with a fervent desire to purchase.

Granularity at the Component Level

Advanced analytics aids in portfolio optimization efforts, in particular machine learning. "Machine learning helps to predict and classify," says Maria. It's essential to get granular at the component level, as the components—parts to build, assemble, or purchase—drive complexity and not the number of variants. For example, a potential candidate to kill can have shared components with other finished goods. This means that the business still needs to support the shared components after killing the low-performing product.

Advanced analytics can show the shared components for teams to understand the impact on profitability of the portfolio optimization efforts. In addition, algorithms can detect similar components—not being the same or identical—showing unexplained price discrepancies and directing to the most competitive suppliers.

What-if analysis[26] extends beyond optimizing the current strategies. It also helps to plan for unexpected and unthinkable scenarios, for companies to adapt swiftly to these changing conditions.

External factors can have a negative impact on well-structured portfolios and can also have a positive impact and uncover opportunities. By anticipating the adverse or shiningly positive conditions, what-if scenarios can provide insights and get companies more prepared to protect and increase their profitability, limiting risks at the same time.

What-if scenarios can also acutely reveal business data, which are challenging to visualize. For example, a new product launch may look like a profitable opportunity on the surface with the expected revenues and associated costs, but what-if-analysis can expose that this new product introduction has a negative impact on the demand for other products (cannibalization). There are also cases where the product by itself is a low performer but creates a significant positive impact on the overall portfolio. Companies also use what-ifs for variance analysis, defining the ideal quantities of the products they need to offer and lowering cost by limiting excess inventory.

Trade Promotion Management

Per Gartner's definition,[27] trade promotion management (TPM) and trade promotion optimization (TPO) are the processes and technologies that consumer goods manufacturers leverage to plan, manage and execute the activities that require collaborative promotional activity from their retail partners.

Like the Bermuda Triangle and the Loch Ness Monster mysteries, TPM effectiveness on sales and profitability remains unknown. Managing trade promotions is becoming more and more important. In its research published in 2020, Deloitte[28] identifies three key shifts:

Shift one: Consumers have low brand loyalty, or they are "brand cheaters" in Maria's terms. There is great price transparency fueled by digitalization, as consumers can get on the internet, to check and compare prices. From the manufacturer's standpoint, this creates more uncertainty about consumer spending on branded products.

Shift two: Retailers are increasing their focus on private label brands, aiming at improving their negotiation power. In addition to this, retailers are bringing exclusive products to differentiate themselves from other retailers.

Shift three: This shift is about how retail formats are evolving. There are new channels, including convenience stores like 7-eleven, gas stations like Shell, ExxonMobil, or BP, and e-commerce. Retail formats are becoming more diverse and smaller, making competition for shelf space fierce within CPGs.

Deloitte states that there are immense opportunities in trade promotion for both manufacturers and retailers. Analytics play a key role in helping CPGs gain insights and shape their offering about promotion design.

Many companies are not tapping into these potential opportunities. Although CPGs are increasing their spend in TPM, only about 20% of companies link trade promotion spending to sell-out success at the point of sale, per Deloitte's study.[29] There is limited transparency about whether specific promotions work. Thirty percent of the surveyed companies indicate that less than a third of promotion events have data about the mechanics and features. This means that CPG companies fail to allocate their budget to profitable promotions.

CPG companies also have limited tracking capabilities to compare forecasts and actuals. Although most of them are comparing results with forecasts based on volume or ROI, they don't have the analytical tools to understand the factors that drive success; the focus is on the differences on volume. "We want to use ROI and consider all the variables," Maria adds.

In its research, Deloitte suggests using analytical tools to fundamentally change this blurry situation with TPM. Per Deloitte, best practices include the following:

- *Create a single repository of all promotions.* Companies need to have a formal way to compare forecasts vs actuals, integrate plan data with customer or account and category teams, and align the data with financial and operations planning.
- *Develop a promotion strategy.* This strategy should allow for budget allocation by customer.
- *Leverage the use of advanced analytics.* Advanced analytics support real-time decision-making based on ROI and strategy.
- *Align and standardize master data.* The alignment and standardization of the data enable companies to have different visualizations of the data and different levels of granularity. "PowerBI is amazing for this," says Maria, and adds "in all cross-functional efforts, there are different interests. PowerBI gives you the flexibility to show each team what they only want to see."
- *What-ifs or scenario simulations.* With scenario planning, companies can build future potential scenarios, assess impact, and define the best parameters.

The execution of these best practices requires that CPG companies work in collaboration with retailers to define together the best mechanics by type of product. The best promotions are those that benefit CPGs and retailers. Everyone is happy.

"Let me give you these best practices in a series of digestible steps, like a recipe from Grandma," says Maria.

"Are you gonna say the secrets too, because cooks provide recipes but don't share those little things that make the difference," says Mike.

"Something even better. After we go through the steps, we are going to implement or cook together. It's not going to be perfect, but it will turn out great," answers Maria.

TPM in 5 Steps
1. Get historical data.
2. Apply analytics to understand demand. "In your toolset, you have regression analysis that is a good option, because we need a causal forecasting technique," says Maria.
3. Decompose total sales into base volume, promotional lift, and cannibalization in your product portfolio.
4. Forecast demand. "As we are using regression analysis or another causal forecasting technique, the model is going to be predictive," clarifies Maria.
5. Run what-if analysis to simulate different scenarios and optimize promotion.

"Team, this takes us to explore further demand sensing and demand shaping. Let's go!" says Maria.

Demand Sensing

Demand sensing refers to a functionality built into some planning applications that allows us to update the forecast when new information becomes available. "Did you watch any of the soccer matches in the FIFA World Cup?" asks Maria, who is originally from Uruguay, where soccer is a popular sport.

"You are talking to a guy from Mexico. Soccer is in my blood," says Ruben.

"I was born in Puerto Rico," says Ernesto.

"I was born and raised in Pennsylvania but watched several games anyway," says Mike.

"So, you are probably familiar with those kinds of predictions or forecasts that people do before the championship starts," says Maria. "Sarah, Jon, Amy, Tim, there are grids with all the matches in the first round and people need to guess the scores. They have time until the first match starts," explains Maria.

"What does this have to do with demand sensing? Are we in a break or kind of soccer class?" asks Jon with his usual irony.

"Here comes demand sensing. Let's say that you guys turned in the complete grid. With demand sensing, you have the opportunity to revise your predictions after you see part of the actual game, for example, at the end of the first half."

This example shows that demand sensing captures real-time market events and adjusts the forecast accordingly. In addition to the traditional order and shipments data and POS sales, the data can come from different sources including social media, the weather forecast, or Nielsen data. If Alex's Snacks estimates sales of a new flavor of chips with onions in 2 million units for the quarter, and the company has already sold 1 million units in the first month, Alex's Snacks has the opportunity to adjust the forecast and share it with the supply chain to take action. This is what demand sensing allows for.

Demand Shaping

Demand shaping is a type of what-if analysis[30] among others like profitable to promise, procurement strategies, fuel cost changes, use of outsourcers, and evaluation of working capital policy. What-ifs allows companies to assess different alternatives, policies, and tactics to boost profits and revenues, to lower costs, and to improve working capital while considering the committed service levels and supply chain constraints.

A fundamental aspect for these simulations to be effective is adopting a holistic approach or multi-dimension analysis,[31] anticipating what's coming, either positive or negative. Demand shaping, in particular, includes analyses of pricing, promotions, and customer deals. It shows the possibilities and the impact of each of these possibilities for teams to maximize the impact on revenues, profits, and volume.

Step-by-step best practices are as follows:

Step 1—Enter the data to run the what-if analysis in the system. For example, to run a promotion, data includes the expected demand uplift, pricing and cost. Teams can do this by making changes to a demand plan or through a scenario simulation.

Step 2—Re-create the plan based on the constraints and optimize the supply plan. Optimization can be to achieve maximum profit or minimum cost, considering the potential increase in sales because of the promotion. The team can evaluate multiple best scenarios.

Step 3—The re-optimized plan defines the impact on overall financial performance, product, and campaign profitability. There are different data visualizations, including P&L by product and by business unit and impact of the campaign against the baseline.

Step 4—Define key drivers of performance.

"We have our root cause analysis toolset," says Amy.

"Excellent point," says Maria, "we need to understand the reasons. For example, we can have stockouts or material shortages, preventing us from fulfilling the potential additional demand because of outdated inventory policy."

"Can we also use Design for Experiment (DOE) or Taguchi?" asks Sarah.

"Someone has been reading non-stop," says Ruben.

"Perfect suggestions about DOE or Taguchi, Sarah! They are also in our lean six sigma toolkit," says Maria.

Step 5—If there are several complex scenarios, teams have the possibility to ask the system to select the best option based on total profit impact. In some situations, it's possible to fulfill a campaign up to a point where there is a big change in cost because of the need to find additional capacity, for example.

Step 6—Teams have access to the opportunity or marginal values for the discovery of further opportunity. These values represent the net income impact of selling an additional unit of product or adding a unit of capacity.

"Proud of you, team! See how far you have come!" says Maria. "Before moving to the supply side, we have Chips in Action."

Chips in Action

1. Obtain the data for forecasting. If gathering the data represents a challenge, you may use some of the tools like process mapping, C&E, and FMEA to tackle it.
   Please also bear in mind the MSA or measurement system analysis, a statistical study that defines whether the measurement system is providing reliable data for decision-making. A Gage R&R study is a statistical study for continuous data and Attribute Agreement Analysis for discrete data.
2. Define the goals for forecasting.
3. Based on the goals and the criteria that can be in a Pugh matrix, select your forecasting technique. Remember that what matters is the net impact on the financials and not the highest forecasting accuracy. Show me the profit—and the money!
4. Run what-if scenarios and demand shaping to understand the different possibilities and the impact on profit, revenues, cost, and working capital.
5. Get ready to show different visualizations of the data quickly. You can use PowerBI to accomplish this.

Supply in Depth with a Holistic Approach, Considering the Impact on the Financial Statements

> *"Learning how to keep track of inventory and cash flow and creating an income statement and a balance sheet are great skills to learn for managing existing businesses."*
>
> Steve Blank

Inventory Management and Optimization

"The portal from the demand side to the supply side is inventory. The feed or input is the demand plan, and the output is the supply plan. Inventory connects both demand and supply," says Maria. Welcome to this new dimension. Like in the Star Trek series, we have the mission of exploring strange new worlds, seeking out new life and new civilizations, and boldly going where no one has gone before.

"We know inventory all too well," says Mike.

"I know you all do, Mike, but here we consider a different angle with finance, too," explains Maria. "I hope you don't mind if we cover the basics first to then get to the meat or to the hot chips with the following:

- Understand inventory from supply chain and finance perspective
- How to conduct an effective inventory analysis—quantity and dollar ($) amount
- Make the best decisions on inventory: what to buy, how much, and when, to optimize inventory by quantifying impact
- Calculate runout dates, liabilities, and other analyses for a successful change management
- Top metrics to track based on SCOR DS framework"

Supply chain and finance together, forever and ever

Maria starts, as she says, with the basics. From an accounting and finance perspective, inventory consists of three distinct groups corresponding to the three production stages.

Inventory

1. *Raw materials and components*—Within raw materials, it's common to hear ingredients and packaging. At Alex's Snacks, the ingredients are potatoes, corn, flour, oil, salt, and seasonings, just like the ingredients that Martha Stewart or Guy Fieri use in their recipes.

Packaging plays a Pivotal role in CPG companies. Yes, the P for pivotal is capitalized because it's important. E-commerce, sustainability, and digitization drive innovation in packaging. Packaging protects the product, provides information, and enhances customer experience. Just as people do judge a book by its cover, consumers do judge a product by its packaging.

There are three levels of packaging: 1. primary packaging is the one that protects and touches the product. 2. secondary packaging is the packaging of the finished goods like cases, and includes boxes or containers that have a specific quantity of primary packaging. 3. tertiary packaging includes pallets and shipping containers for storing and warehousing.

From an operational viewpoint, this is the world of direct materials—all that goes into the products to sell. "Boom!... I mean BOM!" says Maria, "In the bill of material (BOM) is where you can find all that's needed to have the incredible delicious chips. For some companies, BOMs have the labor component as well."

"You, guys, have access to the BOMs in SAP, right?" Maria asks.

"We do," answers Sarah. "The issue is that many BOMs are not updated."

"We are better investigators than those in CSI trying to pull the different pieces together," says Ruben.

"In other words, Maria, we don't have a single source of truth," Ernesto indicates.

"Do you have an estimate of the impact on the financial statements that the BOM errors are causing?" asks Maria.

"I think you, Amy, are working on that," says Ernesto.

"Yes, I am. We've had some instances where we bought items that we don't need."

"I can help you to quantify the impact; if it is significant, I'd suggest following the DMAIC approach," Maria indicates.

"It's going to be helpful," says Amy.

"We'll go over components, WIP, and finished goods and we'll get right into that" says Maria.

Like with raw materials, companies use components to create finished goods. The difference between raw materials and components is that components remain recognizable in the finished goods, like screws, nuts, and bolts. In contrast, raw materials are unrecognizable in finished goods. "We don't see the oil or seasoning in the chips," says Maria.

2. WIP—WIP or *Work in Progress refers to items in the production process.* They include raw materials or components, labor, overhead, and packing materials. There have been some advances, but they are not yet finished goods. For example, the team needs to bag the tortilla chips or label the cases of the original chips.

BOMs can also have WIPs as an ingredient. This situation can happen when the product itself is the same, but the packaging is different.

3. *Finished goods*—These are items ready to sell and to hit the company's financials.

Make to Stock (MTS) and Make to Order (MTO)

Alex's Snacks relies on co-manufacturers for production, having the company make-to-stock as the main manufacturing workflow.

Make-to-Stock (MTS) means producing the items in anticipation of the demand. This anticipation of demand is the forecast for Alex's Snacks—a prediction of future demand. This is a push supply chain strategy.

The challenge of this strategy is that it is based on a forecast. The forecast can use the "peanut butter approach," treating most of the SKUs and decision points the same, although there are different demand patterns as companies do the forecast at aggregate levels with no granularity of how much the individual SKU sells at each store.

Alex's Snacks also follows a Make-to-Order (MTO) manufacturing workflow. In this case, the company produces based on actual customer demand, on receiving actual orders from individual customers. This is a pull supply chain strategy.

The advantages of a pull system are customization and lower inventories. The disadvantages are longer lead times than those in MTS if the raw materials and components are not readily available because the customer still needs to wait for the manufacturing to take place.

Most CPG companies have a hybrid approach with decoupling inventory in the form of raw materials, components, WIP, and indirect materials like MRO that support production.

"Decoupling inventory? What's it?" Tim asks.

"It is inventory set aside to cover for low-stock situations or breakdowns in a different production stage," Maria says.

"Gotta you! Decoupling inventory is to keep things flowing," says Tim.

While safety stock acts as a buffer against changes and disruptions by external sources like customer demand, decoupling inventory is a cushion against changes caused by production, the internal demand.

"Reassuming the BOM issues that we were talking about, let's dive into the impact of inventory on the financial statements, including the different transactions, such as purchase of raw materials, production, and when we sell our potato and tortilla chips!" says Maria.

Impact on Financials

"Where can I see the inventory that Alex's Snacks has?" asks Maria.

"Right here; point 4 in assets, on the balance sheet," answers a chest-beating Amy.

"That's right, Amy!" says Maria.

"And for raw materials, there's more in inventory at the co-manufacturers that is off the books but that we are still liable for," adds Tim.

"For finished goods we have lots at the distribution centers," says Mike.

"If you fix that, it would be a miracle," adds Jon.

"Challenging, right?" Maria says. "We now know what information to look for to define our baseline for inventory. We also want to have a solid measurement system to determine the improvements, our dear MSA."

Cost Accounting
"Like the philosopher's stone in the Harry Potter books, cost accounting is core to manufacturing, accounting (duh), and finance," says Maria. "We need to understand cost accounting to make our savings real in front of finance's eyes, real like holding and feeling the dollar bills in their hands," she adds.

"Like how that sounds," says Ernesto.

"What's cost accounting?" asks Mike, like Inspector Jacques Clouseau in *The Pink Panther*.

"Is that related to ABC costing? I had that at school," says Tim, "it was like a mumbo jumbo with all the numbers."

"More numbers?" asks Jon, getting up off his chair and walking towards a big window to peek at the building's surroundings. He comes back to his seat and asks, "We saw tons of numbers with the financial statements, what is this cost accounting about now?"

"Jon," says Maria, pausing for a few suspenseful seconds like in a Miss America pageant before the winner is announced. "Your question is the perfect fit."

Financial Accounting is for external parties and follows defined standards, while cost accounting[32] (totally optional) is for the internal team to identify areas for cost reduction and containment and advise on pricing to ensure profitability. It assists a combo of three types of decisions: budget, product pricing, and strategy.

"Budget, the Halloween word, right Maria?" says Ruben, using the same terms as Maria.

"Yes, the inputs from cost accounting to the budgeting process are phenomenal. The deeper we understand the actual variable and fixed costs, the better the future projections and allocation to the different product lines," explains Maria.

As its name denotes, cost accounting is all about costs: materials, labor, and overhead. Each of these three buckets contains sub-buckets with direct and indirect costs.

Direct costs—materials and labor—are clear. If it is in the BOM, it is a direct material. The labor cost of workers involved in production is direct. Indirect material and indirect labor costs are like overhead expenses. The cost sheet includes direct materials only. Likewise, indirect labor is not a labor expense.

"A lot of theory, Maria," says Jon. "How does this help us? Are the next Chips for Thought or Chips in Action on how to classify them? That's a non-sense exercise for experienced supply chain and procurement people, as we are."

"Jon, again, a very good point to explain how cost accounting can help in a more actionable way," replies Maria.

Understanding the breakdown of materials, labor, and overhead costs helps procurement in their negotiations. The labor cost in China is lower than in the US. For example, if a product has 80% of materials, 10% labor, and 10% overhead, there are not many advantages to sourcing it from China, as the labor component is low. This is considering that China and the US are using the same materials. Knowing the cost breakdown also aids procurement professionals working together with suppliers in finding opportunities for cost reduction.

Standard Costing and Activity-based (ABC) Costing

"Standard costing, ABC costing, that's it!" says Tim, reading these words on the large whiteboard in the conference room.

"I hope I will be able to turn around those memories from school, Tim," replies Maria with a smile and adds, "I don't want you to see cost accounting and run like my youngest daughter, Rachel, when it's time for a bath."

There are various approaches within cost accounting to analyze production expenses, with standard costing and activity-based costing the most well-known. Standard costing is an estimate of costs considering the efficient use of resources like ingredients, packaging, and labor, in typical conditions. There is then a comparison between the estimates and actuals.

The difference between actuals and standards is powerful. It can reveal the need for process improvement or poor performance from suppliers. "When things go wrong, it's the supplier's fault; if things go great, it's us, of course," says Maria.

"Can't agree more," endorses Ruben, and adds, "If we need to improve the processes, we do all the trees like the fault tree analysis, right?"

"Yes, you can use the tools from your toolset to find out the root causes and wipe out the issues," replies Maria.

ABC costing identifies the cost drivers or activities and allocates overhead costs based on them. This means that the activity that builds more cost or uses more resources gets a bigger portion of the overhead. This is like a fairer option in comparison to the classical allocation of overhead that is based on value or quantity. But, ABC is challenging and costly to implement.

ABC and job costing are for companies with diversified products where labor is a major component, like in big accounting or legal firms. Standard costing and process costing apply to companies that mass-produce standardized products, like Alex's Snacks do. The assumption here is that each item is like the others, allowing us to allocate indirect production costs evenly across the company's output.

Inventory Transactions (Normal and Standard Accounting)

There are different transactions related to inventory that have an impact on the financial statements. Quantifying the impact is critical for supply chain and procurement teams to communicate with finance and with the business.

"This is a high-level summary of a few accounting rules," introduces Maria.

When assets and losses increase, it's a debit; otherwise, it's a credit.

When liabilities, equity, and gains increase, it's a credit; otherwise, it's a debit.

Debits go to the left side and credits go to the right side.

"We walk through the transactions considering the normal accounting system as well as the standard accounting system," explains Maria. In the normal accounting system, companies allocate the overhead cost based on a predetermined overhead rate (estimate) multiplied by the actual activity level or volume.

Normal Accounting

Overhead cost allocation = Predetermined overhead rate X actual activity level

where predetermined overhead rate is based on estimated cost and estimated activity levels.

In a standard accounting system, the overhead allocation is the standard rate multiplied by the standard activity level. The standard rate is the same as that in the normal accounting.

Standard Accounting

Standard—standard rate X standard activity level

Where standard rate = predetermined overhead rate in normal accounting system

"There are standards for quantity and for rates," says Maria and adds, "to compare them against actuals, we use this formula." She writes the formula with a black marker on the whiteboard.

Actual overhead = Actual overhead rate X actual activity level

Maria continues her explanation with, "To get the actual overhead rate we need to wait until the period is over."

Standard costing[33] uses the same approach to track cost during the manufacturing processes, not limiting it to overhead only. For example, it records purchases at a standard cost, issues materials into production at standard quantity and cost, direct labor at standard hours and rate, and overhead at standard rates and activity levels.

"The key here is to remember that there are standards for quantity and for cost or price. Procurement, in the level of importance for your savings calculations, this is close to the height of Mount Everest's peak, the highest altitude above sea level," says Maria.

The cost savings calculations need to show the quantity and cost of the following:
- *Baseline*—If the company uses standard costing, procurement needs to show the standard quantity and standard rate. If it does not, procurement needs to define a baseline quantity and a baseline cost along with the reasons for choosing that baseline. The MSA in the lean six sigma toolkit is helpful to ensure that the measurement system is reliable.
- *Estimates*—When showing projected annual cost reduction, procurement should identify and state quantity and cost separately. It needs to be super, super, super clear: baseline cost, new cost, baseline quantity, new quantity.
- *Actuals*—Procurement shines with a robust plan for data collection of the actual quantities and cost to compare against the savings estimate and baseline.

"That's it! We are ready to see the transactions," Maria says.

Raw Material Purchases

When a business buys raw materials, it increases inventory and decreases cash if there are no payment terms with the supplier, or if there are payment terms, it increases Accounts payable. Below are the journal entries in both situations:

1. Alex's Snacks buys raw materials for $100 paying at the moment of delivery, or COD (cash on delivery). That means no credit terms.

Raw Materials		$ 100	
	Cash and cash equivalents		$ 100

Raw materials (inventory, asset) go up and Cash and cash equivalents (asset) go down by the same dollar amount. The change is in the composition of the assets that a less liquid asset (raw materials) comes in while a more liquid asset (cash) comes out. The net effect on assets is zero, the same as with the balance sheet. No impact on the P&L.

2. Alex's Snacks buys raw materials for $100 with terms of net 60 days.

The journal entry with the goods receipt is below.

Raw Materials		$ 100	
	Accounts payable		$ 100

Raw materials (inventory, asset) go up and Accounts payable (liabilities) go up by the same dollar amount. The company owes that money to suppliers. Assets go up because of the increase in raw materials that is inventory and liabilities go up too because the company has a new liability with the raw material supplier. Terms are net 60 days. The net effect on assets is an increase by $100 at the same time that this creates a liability for the same amount. The net effect on the balance sheet is zero; it changes the composition. No effect on the P&L.

"Are these two situations the same considering the impact?" asks Maria, "Would you have any preferences?"

"How about the cash flow statement? We haven't talked about the impact on it. With net 60 we are better off," says Amy.

"Terrific! Yeah, the second situation is better. It's true that we are incurring a liability, but during the 60-day period, the company can use the funds. Alex's Snacks may decide to invest to get interest." This is what Robert Kiyosaki,[34] author of *Rich Dad, Poor Dad*, a New York times best-seller for years, would classify as good debt.

If the business follows standard costing, the journal entry would be as follows:

Raw materials inventory		$ 80	
Materials Price Variance		$ 20	
	Accounts payable		$ 100

Actual quantity purchased = 100
Standard quantity = 130
Actual price =$1
Standard price=$0.80

Raw materials inventory equals actual quantity times standard cost. This is 100 multiplied by $0.80. Accounts payable is actual quantity purchased by actual price. Plugging in the data, it is 100 (actual quantity) times $1 (actual cost) to equal $100.

Raw materials inventory = actual quantity purchased X standard cost

Raw materials inventory = 100 X $0.80 = $80

Accounts payable = actual quantity purchased X actual cost (this is like in the normal accounting system)

The formula for *material price* variance is actual price minus standard price times actual quantity. This is $1 (actual price) minus $0.80 (standard price) multiplied by 100 (actual quantity) equals $20.

Price variance = (actual price - standard price) X actual quantity

Price variance = ($1-$0.80) X 100 = $20

This is an unfavorable variance.

Interest
3. Alex's Snacks takes advantage of the 60-day period and invests to obtain interest.

Cash and cash equivalents		$ 1	
	Interest income		$ 1

Cash (asset) goes up in the balance sheet, and interest income also goes up in the P&L. What a wonderful world!

In the first situation, when Alex's Snacks pays COD, the company may not have enough funds and has no other choice than to borrow money. In that case, there is a loss. The journal entry is as follows:

4. Alex's Snacks borrows funds to pay COD, paying interest.

Interest expense		$ 1	
	Cash and cash equivalents		$ 1

Cash (asset) goes down in the balance sheet, and interest expense goes up in the P&L. "We can hear Finance screaming!" teases Maria.

"How about the situation of buying inventory that is not needed? It's not our fault that the BOMs are not updated," indicates Tim.

Inventory write-off
"Can we use that inventory for something else? Or to make a minor change to be able to use it? Or can another company use it even at a discounted price? We want to ask these questions, so we don't have a loss in the best-case scenario, or we minimize the loss. We need to find creative solutions," explains Maria. The journal entry for write-offs or losses would be:

5. Alex's Snacks writes off inventory that the company can't use (damaged, obsolete, etc.) for $500.

Inventory write-off		$ 500	
	Inventory		$ 500

Inventory write-off Is an expense account in the P&L. It shows the loss that the company incurs because of inventory that they can't use. The inventory account is an asset account on the balance sheet that goes down. The impact on the financial statements is a loss in the P&L and a decrease in assets and in equity (retained earnings) in the balance sheet. There is no effect on the cash flow statement.

"How about if the company is still not certain about the loss and keeps the inventory for some time while looking for a solution? Where do we see the storage cost for keeping this inventory?" asks Amy.

"This is a good point, Amy, because the inventory keeps generating costs while we keep it. The storage cost is part of COGS when we have raw materials and WIP. If we have finished goods that remain unsold, the storage cost goes into operating expenses (OPEX). You can see why we want high inventory turns. We want to make money with it, and the faster, the better; otherwise, we feed Ernesto's profit Pac-Man," says Maria.

"Are maintenance, repair, and operating items included in COGS or OPEX?" asks Sarah.

"MRO is a broad category. If they are part of the production or factory overhead, they belong to COGS. If they are expenses like janitorial supplies for the office buildings, they belong to OPEX. MRO may also include packaging. I've seen packaging as direct materials (inventory) in some companies and as indirect in others. Either direct or indirect, the packaging related to production is part of COGS in the income statement," responds Maria.

"How about freight?" asks Mike.

"Freight-in or inbound shipping cost goes to COGS while freight-out or outbound shipping cost belongs to OPEX," answers Maria.

"And PPE—personal protection equipment—does it go to COGS or fixed assets?" asks Amy.

"When we have uniforms or other clothing that we use in production, we include them in COGS," explains Maria.

Related to assets and expenses—income statement vs. balance sheet—in December 2001 there was a scandal with Enron,[35] an innovative energy company considered a darling of Wall Street investors with $63.4 billion in assets. The company filed Chapter 11, becoming the largest bankruptcy in US history. The CEO, CFO, and other executives went to prison for fraud and other offenses. Arthur Andersen, the company's auditor, ceased doing business. Enron was capitalizing expenses. This means that Enron was including expenses—that should have gone in the P&L—as assets on the balance sheet.

It's common to use purchase orders to buy raw materials and components. POs don't have a direct impact on the financial statements because there are no journal entries. But, POs can have an indirect effect on the financials because finance may use the open POs to create accruals. Accruals on open POs[36] has become an essential task, as the expenses are only posted in the system along with the goods received. Accruals are reserves to cover for the commitments a company has with its POs.

Raw Materials Receipts with No Invoice

When the supplier delivers the raw materials is when the inventory slams or hits the books. The journal entries for goods receipts are those stated in #1 if COD or #2 if the company has terms with the raw material supplier.

There are situations where the supplier delivers the raw materials without an invoice. Maria remembers that on more than one occasion suppliers sent the invoices months later. This represents a surprise to finance, and finance doesn't like surprises.

To avoid these sour-flavored situations, some companies use accounts like goods received not invoiced, or stock received not invoiced against accrued payables or accrued purchase orders. The journal entries are as follows:

6. Alex's Snacks receives raw materials for $200 without the invoice.

Goods received not invoiced		$ 200	
	Accrued payables		$ 200

Goods received not invoiced is an inventory account within assets on the balance sheet because the company has inventory that they can use. Accrued payables is a liability that indicates that Alex's Snacks owes that amount and that the invoice is on its way.

When the invoice to pay arrives, finance doesn't like it either, but they don't have the sour-flavored surprise and can keep walking and breathing normally. The journal entries are like those below:

7. The raw materials supplier sends the invoice for $200 four months after the delivery of the goods.

Accrued payables		$ 200	
	Accounts payable		$ 200
Raw materials		$ 200	
	Goods received not invoiced		$ 200

As with the invoice, the company has the regular Accounts payable, and the entries show a debit to Accrued payables. For being a liability, Accounts payable goes up on the credit side. Alex's Snacks has the normal raw materials—invoiced physical goods—on the debit side and goods received not invoiced goes down.

Considering the journal entries of #6 and #7 together, goods received not invoiced and Accrued payables cancel each other out.

~~Goods received not invoiced~~		~~$ 200~~	
	~~Accrued payables~~		~~$ 200~~

~~Accrued payables~~		~~$ 200~~	
	Accounts payable		$ 200
Raw materials		$ 200	
	~~Goods received not invoiced~~		~~$ 200~~

Payment to Raw Material Suppliers

The supplier bill is there. It's due. Finance and Accounts payable grudgingly open their wallet to pay. It feels like holding all the hot potato chips in your mouth without having a drop of water. Yes, it's hard to swallow. The journal entry is as follows:

8. Alex's Snacks pays the raw material supplier invoice for $200.

Accounts payables		$ 200	
	Cash and cash equivalents		$ 200

As Alex's Snacks cancels the liability with the raw material supplier, there is a debit for $200 of Accounts payable. The entry includes a credit to Cash and cash equivalents (asset account) as funds are coming out. The total assets on the balance sheet go down by $200, with no impact on the income statement. The cash flow statement receives a negative impact of $200.

WIP Creation

During production, there can be WIP that is work in progress or in process. These WIP are not yet finished goods, but they have added value. Examples of journal entries are below:

9. The team uses some of the ingredients to make the crunchy chips for $100.

WIP		$ 100	
	Ingredients		$ 100

As WIP is inventory, an asset, the creation of WIP goes as a debit on the left side. Ingredients—also inventory—go as a credit when the team uses them in the production process. The impact on the balance sheet is a change in the composition of the assets. No impact on the income statement or the cash flow statement.

If standard costing, the journal entry to record consumption at standard cost is as follows:

WIP		$ 104	
	Materials Quantity Variance		$ 24
	Raw Materials Inventory		$ 80

Actual quantity purchased = 100
Standard quantity = 130
Actual price =$1
Standard price=$0.80

WIP equals standard quantity times standard cost. This is 130 multiplied by $0.80. Raw materials inventory equals actual quantity times standard cost. This is 100 multiplied by $0.80.

Raw materials inventory = actual quantity purchased X standard cost

Raw materials inventory = 100 X $0.80 = $80

The formula for material quantity variance is actual quantity minus standard quantity times standard price. This is 100 (actual quantity) minus 130 (standard quantity) multiplied by $0.80 (standard price), which equals -$24. This is an unfavorable variance.

To close materials price and quantity variance to COGs, the journal entry is the following:

Material's Quantity Variance		$ 24	
	Material's Price Variance		$ 20
	Cost of Goods Sold		$ 4

The journal entry to reflect the work of the team in creating WIP is as follows:

10. The teams work diligently in the creation of WIP for $300. The team then uses WIP for the repacks.

WIP		$ 300	
	Wages payable		$ 300

Like when the team uses the ingredients, WIP goes up with the added work. Wages payable goes up too, a liability for the business. The impact is an increase of $300 on the balance sheet. No effects on P&L or cash flows. When the company pays the team and cancels its obligation, the cash flow statement will show a negative amount of $300.

If the company uses standard costing, the entries are those below.

WIP		$ 416	
	Direct labor rate variance		$ 20
	Direct labor efficiency variance		$ 96
	Accounts payable		$ 300

Actual labor hours = 10
Standard quantity = 13
Actual rate = $30 an hour
Standard rate = $32 an hour

Accounts payable is the actual number of hours multiplied by the actual hourly rate. This is 10 hours (actual number of hours) times $30 (actual hourly rate) equals $300.

Accounts payable = actual number of hours X actual hourly rate (this is like in the normal accounting system)

The formula for *direct labor rate variance* is actual hourly rate minus standard rate times actual number of hours. This is $30 (actual hourly rate) minus $32 (standard hourly rate) multiplied by 10 (actual number of hours) equals $20.

Direct labor rate variance = (actual labor rate - standard labor rate) X actual number of hours

Direct labor rate variance = ($30-$32) X 10 = -$20

This is a favorable variance.

The formula for direct labor efficiency variance is actual labor hours minus standard labor hours times the standard hourly rate. This is 10 (actual number of hours) minus 13 (standard number of hours) multiplied by $32 (standard hourly rate) equals $96.

Direct labor efficiency variance = (actual number of hours - standard number of hours) X standard hourly rate

Direct labor efficiency variance = (10-13) X $32 = -$96

This is also a favorable variance.

To close direct labor rate and efficiency variances to COGS, the journal entry is as follows:

Direct labor rate variance		$ 20	
Direct labor efficiency variance		$ 96	
	COGS		$ 116

11. The company applies to WIP the variable overhead of $250, paying cash.

WIP		$ 250	
	Cash		$ 250

WIP goes up with the added variable overhead, and cash goes down. The impact on the balance sheet is a change in the composition. WIP replaces the most liquid asset by $250. No effects on the P&L. The Cash flow statement shows a reduction of cash by $250.

Following a standard cost approach, the journal entry is as follows:

WIP		$ 260	
Variable overhead price variance		$ 50	
	Variable overhead efficiency variance		$ 60
	Cash		$ 250

Actual variable overhead hours = 10
Standard quantity variable overhead = 13
Actual rate = $25 an hour
Standard rate = $20 an hour

Cash is actual number of hours by actual rate per hour. This is 10 hours (actual number of hours) times $25 (actual hourly rate) equals $250.

Cash = actual variable overhead number of hours X actual hourly rate (same as in the normal accounting system)

The formula for variable overhead price variance is actual rate minus standard rate times actual number of hours. This is $25 (actual hourly rate) minus $20 (standard variable overhead rate) multiplied by 10 (actual number of hours) equals $50.

$$\text{Variable overhead price variance} = (\text{actual variable overhead rate} - \text{standard rate}) \times \text{actual number of hours}$$

$$\text{Variable overhead price variance} = (\$25-\$20) \times 10 = \$50$$

This is an unfavorable variance.

The formula for variable overhead efficiency variance is actual variable overhead hours minus standard number of hours times the standard hourly rate. This is 10 (actual number of hours) minus 13 (standard number of hours) multiplied by $20 (standard variable overhead hourly rate) equals -$60.

$$\text{Variable overhead efficiency variance} = (\text{actual number of hours} - \text{standard number of hours}) \times \text{standard hourly rate}$$

Variable overhead efficiency variance = (10-13) X $20 = -$60

This is a favorable variance.

To close variable overhead price and efficiency variances to COGS, the journal entry is as follows:

Variable overhead efficiency variance		$ 60	
	Variable overhead price variance		$ 50
	COGS		$ 10

Finished Goods—Assemblies

This is an exciting moment—like when the egg hatches, when the smell of cinnamon wafts sweetly from the freshly baked cake, or when the writer publishes a book. It's a finished good! The potato and tortilla chips are ready to sell! This is a transformation journey with the ingredients and packaging that manufacturing makes possible. The journal entries to show the birth of finished goods are like those of WIP, but instead of debiting WIP, we debit Finished Goods. Finished Goods can also have WIP, and in that case, we debit Finished Goods and credit WIP.

12. Alex's Snacks repacks WIPs into smaller sizes for $300.

Finished Goods		$ 300	
	WIP		$ 200
	Film		$ 50
	Wages Payable		$ 50

Finished goods go up, while WIP goes down when teams use it. Film (inventory) that is packaging goes down, and wages payable shows the liability created because of the workers' labor.

"If the company uses standard costing, we have the entries with standard quantity and price, and the resulting variances when comparing standards against actuals. With this, we complete finished goods and we're ready to make some money selling them. Hooray, team!" says Maria.

Finished Goods Sales

"The masterpiece of procurement shows in COGS, but this masterpiece, or the lower cost, may not appear instantly in the income statement because of existing inventory and the valuation method used," explains Maria and adds, "when you guys talk to finance, you are better prepared if you understand the valuation method, as in addition to how much the savings are, finance wants to know when they see the awesomeness of your work on the P&L."

The most common four valuation methods[37] are:
1. First in, first out (FIFO)—FIFO considers that the company sells the oldest products first. When there is inflation, companies applying this valuation method show lower COGS and higher profitability.

2. Last in, first out (LIFO)—LIFO is the opposite of FIFO. LIFO considers that the company sells the most recent inventory first. In inflationary times, LIFO shows higher COGS and lower profit margins.
3. Weighted cost average—With this method, the company averages the existing with the new inventory.
4. Specific identification—The specific identification that involves monitoring the cost of each individual unit is not common in CPG companies. Companies that sell high-priced individual products may apply it.

There are two journal entries: 1. to reflect the sale of finished goods and 2. to record the cost of the finished goods sold. Here is where the valuation methods come into play.

13. Alex's Snacks sells to Walmart $1000 in crunchy potato chips.

Accounts receivable		$ 1,000	
	Sales		$ 1,000
COGS		$ 500	
	Inventory		$ 500

Sales increases by $1000 in the income statement or P&L. As Walmart doesn't pay with the receipt of the goods, accounts receivable (assets) go up on the balance sheet.

There is a debit to COGS and a credit to inventory to show the cost of this sale to Walmart and to update the inventory levels at Alex's Snacks.

The difference between these couple of entries shows the gross profit of the sale that is $500, representing a 50% gross margin.

"How are we feeling?" asks Maria.

"Good," answers Amy.

"Lots of stuff, Maria," says Mike.

"I feel I understand accounting and standard costs better now," says Tim.

"Thank you so much, Tim, for your comments. I'd like to add that it takes more than once to fully grasp all the concepts and applications, so it's good that you and everyone on the team revisits this periodically. Next time it's going to be easier," says Maria.

"Let's have some chips to move on to inventory transfers."

Inventory Transfers

Inventory transfers occur when a company moves inventory from one location to another. The company owns the inventory, and the transfer transaction's purpose is to faithfully reflect where the inventory is.

An inventory transfer transaction doesn't have a direct impact on the financial statements because it means a change in location only. But, it could have an indirect impact if the inventory is not shown in the proper location, and the company needs to buy or produce more inventory to meet a deadline for a customer, for example. A good practice for all things warehousing is 5S in the lean six sigma toolkit. "Remember when you helped me with my purse?" asks Maria. "That's 5S."

Finished Goods Returns

Customers may return the products to the company. "To show the returns, we flip the journal entries corresponding to the sale. It's like that," says Maria, snapping her fingers.

14. Walmart returns $300 in sales to Alex's Snacks.

Sales		$ 300	
	Accounts receivable		$ 300
Inventory		$ 150	
	COGS		$ 150

"That's easy for finance and accounting. For operations, returns can be a nightmare," adds Mike.

"Give a hand for all of you, team, for going over the inventory transactions. I know it's challenging but worth it!" Maria says.

"Bring it on, finance," says Ruben.

"Maria, nice job!" says Ernesto. "I have other meetings but will catch up later. Team, keep up the great work!"

"Thank you so much, Ernesto and team!" says Maria; "Let's jump on ABC analysis!"

ABC Analysis

"To avoid confusion, this is not the same ABC as that in cost accounting," clarifies Maria.

"Thank heavens!" says Tim.

"Don't count your chickens before they hatch," advises Jon, "how do you know this ABC is better?"

"You're right, Jon," replies Tim, "This can be worse."

"Guys, I know that it's not easy to analyze the data and get insights to take action, but this is also an opportunity. It depends on your perspective," encourages Maria. Tony Robbins indicates that problems are not stumbling blocks; they are stepping stones to something better if companies and people use problems to their advantage.

"That's a fairy tale, Maria. I'm a busy man who works his butt off day in and day out. With all of these novelty ideas, I can't work; and now I need to stay at the office longer to come back home and hear my wife whining because I'm late. Can you fix that?" Jon challenges Maria while the other team members remain quiet.

"Like other trainings, this training requires a time investment upfront. The frameworks and toolsets allow you to prioritize based on the impact on the financials and you can perform better. The training helps you," Maria states.

Precisely, ABC analysis helps to prioritize the inventory to monitor. ABC analysis[38] consists of determining the financial importance of the inventory items, either based on sales (monetary units) or volume. Like in the DMAIC framework and other classifications, the objective is to focus efforts on the items with the highest impact. A items are those that rank first considering the impact, B items follow A items, and C items rank after A and B items. Other classifications include more item categories, extending the item ranking to D and to E and F groups.

These item groups or categories are based on the Pareto Principle, which states that 80% of the impact on the financials comes from 20% of the items. "You have the Pareto analysis in your toolkit," says Maria. Considering this, a typical company like Alex's Snacks has a few selected A items that they treasure like gold, a larger group of B items that are important too but not as important as the A items, and lots of C items that are the vast majority of items.

Type	Importance	% of Total Inventory	Annual Consumption Value	Controls	Records
Class A	High dollar	10-20%	70-80%	Tight	High accuracy
Class B	Medium dollar	30%	15-20%	Medium	Medium
Class C	Low dollar	50%	5%	Basic	Low

Source: Oracle Netsuite[39]

ABC analysis has several advantages. It helps to optimize inventory because companies know the demand and treat the most important items (A items) like kings and queens, reserving room for them in the warehouse to keep these items in good condition, in sufficient quantity, and controlled with the accuracy of a plastic surgeon. Still belonging to the royal family but not in the high ranks of the A items, B items have medium levels of controls and records. The vast majority of items in the kingdom fall in the C category, with basic controls and low accuracy records.

In addition to its positive contribution to inventory optimization, ABC analysis improves inventory forecasting accuracy as teams get closer to the demand of the high sellers and develop a better understanding of it.

"In my notes, I have that sales accuracy is higher for items with high sales," advises Amy.

"Absolutely correct, Amy," says Maria. "Also, knowing the demand, we can manage pricing better and generate more profits for Alex's Snacks. This is another example of why supply chain creates value at the top and bottom lines."

"Talking about the bottom line," says Sarah, "with the ABC classification, we can focus on the price negotiations with the suppliers of the A items."

"Yeah, Sarah, and we can also see if we can consolidate suppliers to leverage volume and obtain better results in our supplier negotiations," adds Tim.

"Great, team! Keep those thoughts coming!" says Maria. "As you can note, the fundamentals of the ABC analysis lie in the categorization of the items and taking action accordingly. If we treat them equally, we are not directing our efforts where the value is."

Treating the items per their category translates into anticipating and following demand to keep the right quantity, ensure warehouse space, and offer higher service levels for the most important items in profitability terms.

"I can feel that you can't wait to put your hands into the inventory data to perform the ABC analysis," says Maria.

Chips in Action

Perform ABC analysis with your inventory. "I have step-by-step instructions to do so," clarifies Maria.

ABC Step-by-Step Instructions

Step 1: State the why. "Surprise, surprise, team, we start with the why," says Maria. There are two key objectives in performing ABC analysis: 1. cost savings opportunities in procurement or spend analysis, and 2. optimize inventory levels considering sales or production.

Step 2: Obtain the spend data and the ordering and holding costs. For the spend data, you may use purchase orders and receipts, as the supplier invoices may not have all the details.

"We want to know the annual spend of each item, but we don't want to stop there," says Maria, and adds, "we want to know about sales and profit margins too."

Step 3: Obtain sales and profit margin data from sales orders and information from finance or cost accounting.

As a cheat sheet, the data for ABC analysis is as follows:
- annual spend by item
- ordering costs
- holding costs
- sales or consumption by item
- profit margin

Step 4: Use the ABC formula to then rank each inventory item based on cost, in descending order, from the highest cost to the lowest. To ensure having all the items, compare against the item master from the ERP system.

ABC formula

Annual usage value = Annual number of items sold or consumed X Cost per item

"Many times, I run three ABC analyses by: 1. cost, 2. units, 3. profit, taking them all the way until the final ABC designation. Then, using cost as a basis, I adjust for any A's that show up in the other two analyses. It allows for better coverage," explains Maria.

Step 5: Determine the impact by dividing the annual spend by each item divided by total annual spend.

% Impact = (annual item cost)/ (annual spend) X 100

Step 6: Categorize items in A, B, and C and take the appropriate actions. Keep in mind to treat A items like kings or queens, B items like the extended royal family, and C items like regular people in the kingdom.

Step 7: Analyze opportunities, including supplier negotiations, issue a Request for Proposal (RFP), etc.

The ABC analysis has the caveat that there is no one way to do it. As inventory may include raw materials that do not have a profit, companies perform ABC based on cost. With finished goods, companies may look at their more profitable items.

"Many times, I run three ABC analysis by: 1. cost, 2. units, 3. profit, taking them all the way until the final ABC designation. Then, using cost as a basis, I adjust for any A's that show up in the other two analysis. It allows for better coverage," explains Maria.

ABC analysis has bright advantages but also dark disadvantages. One of these dark disadvantages is the changes in parameters. For example, some of the B or C items can become A, making the team reassess with all the time and effort that implies. Related to the parameters and categories, team members' judgments play a big role, adding their dose of subjectivity to the analysis.

Other dark disadvantages include stockouts or excess inventory for B and C items, as the focus is on the A items. Both stockouts and excess inventory can have a negative effect on production and on the financials. For instance, when there's excess inventory, the risk of obsolete and aged inventory increases. ABC analysis is not a good fit for new product introductions or seasonality because the focus is on the high-dollar or volume items.

"And it's soooooo time-consuming!" says Mike.

"Can't argue with that," says Maria. "It can be like a vacuum cleaner sucking up resources with the assessments and reassessments, but we can use automation in Excel to have a less resource-intensive analysis and benefit from the greater profits and process efficiencies that ABC brings."

What to buy, what to produce, how much, and when

"This is planning 101 and where the pieces of the puzzle fit together," says Maria. "How do we start?"

After a few seconds an answer comes in the form of a question.

"With the demand?" asks Tim.

"Yes, Tim," answers Maria. "We start with the demand for finished goods, the forecast and techniques that are in the previous chapter of our journey." For the forecast of finished goods—also known as the independent demand—there is a unit of time associated. For example, a company can have a forecast by quarter, month, week, day, etc. "We are using weeks," states Mike.

"Good! What do we want to know next, once we have our demand by week for the barbecue chips—for the sake of the example—if we wanted to know what to produce?" asks Maria.

"The inventory we have on hand," answers Tim, this time without hesitation.

"Perfect! What we already have physically in the warehouse or warehouses and distribution center or centers. What else do we need to know?"

"The work orders to be completed each week," participates Mike.

Considering this conversation between the team at Alex's Snacks and Maria, the data to look for is as follows:
- Demand for finished goods. To simplify, the example here is with the barbecue flavored potato chips, with the demand per week.
- On-hand inventory
- Work orders to be completed each week

Example with Barbecue	Week 1	Week 2	Week 3	Week4
Demand	2,000,000	1,000,000	3,000,000	3,000,000
On Hard Inventory	1,000,000	3,000,000	4,000,000	2,000,000
Work Orders	4,000,000	2,000,000	1,000,000	2,000,000
Projected Inventory	3,000,000	4,000,000	2,000,000	1,000,000

The table shows the projected inventory considering this equation:

$$\text{Projected inventory} = \text{On hand inventory} + \text{Work orders} - \text{Demand}$$

Where

On hand inventory equals beginning inventory or the previous period—in this case, a week—projected inventory. For example, on hand inventory of week two is three million; that is the projected inventory at the end of week one. Likewise, on hand inventory at week three is four million; that is the projected inventory at the end of week two.

By looking at the table with the projected inventory of the barbecue-flavored potato chips, the team can create simulations with different production lot sizes and lead times, and make changes to the production schedule.

"This is for finished goods. For raw materials and packaging, we need to look at the bill of materials (BOMs) to determine the dependent demand. This is known as MRP (Material Resource Planning)."

"Please don't get Sarah started with the BOM issues," says Ruben.

"Yeah, Maria, sometimes I get frustrated," Sarah vents.

"How can you not? We get the heat when we don't have the right materials on time and got others instead that we don't need," Amy says in support.

"And that heat feels more intense than waiting in line for hours to ride the Space Mountain in Disney on Memorial Day!" adds Ruben.

"Understood. If the BOM has errors, we can end up with the wrong materials," agrees Maria. "We see this in the MRP example." Maria shows a table for the flexible packaging of the BBQ chips.

Film Barbecue Cases	Week 1	Week 2	Week 3	Week 4
Demand	2,000,000	1,000,000	3,000,000	3,000,000
Production	1,000,000	500,000	1,000,000	1,500,000
On Hand Inventory	1,000,000	300,000	100,000	100,000
Scheduled receipts	300,000	300,000	1,000,000	1,400,000
Projected Inventory	300,000	100,000	100,000	0

The table shows the demand in the first row. In the S&OP or IBP process, considering the most evolved version of S&OP, the demand is a forecast that companies develop to anticipate customers' orders. It becomes the unconstrained demand. Unconstrained means that all the demand is in there, with no checks against capacity or supply limitations.

The second row in the table, production, is the result of assessing the supply capability, including the review of inventory available, aggregating all sources of supply, and adjusting supply planning constraints to determine the production plan. Production planning[40] consists of modeling the available capacity to optimize workloads while following the changes in demand.

Continuing with the terminology, capacity modeling is an analysis of the output of how many units manufacturing can produce in a certain timeframe, like an hour or a day, by using specific work centers or resources. A common measure is ADR, which stands for average daily rates. In our example, ADR measures how many cases of barbecue chips the co-manufacturers of Alex's Snacks produce each day, on average.

With the master production schedule (MPS) that adds timing to the list of products to manufacture, the team optimizes bottlenecks, resource constraints, and capacity issues. In doing so, the initial unconstrained demand becomes constrained because of the limits on the supply side.

The MPS works in conjunction with *rough cut capacity planning* to determine the requirements of key resources like direct labor and machine time and with *material requirements planning* (MRP) to define the requirements of raw materials and packaging.

"We can see that the other rows in the table show inputs to the MRP," explains Maria. The inputs of an MRP[41] are as follows:
- Master production schedule (MPS)
- On hand inventory information
- Scheduled receipts
- Bill of materials (and routes)

"The MPS goes through only one layer of the BOM, while the MRP goes through all layers to determine every requirement," clarifies Maria and adds, "We can see that BOMs are key inputs for the MRP. If we have them wrong, our planning will be wrong; kind of garbage in, garbage out."

Data visibility and accuracy on the demand, production, BOMs, on-hand inventory, and scheduled receipts allow companies to go through the change management process successfully. It is fundamental to estimate runout dates of old part numbers, potential liabilities, and the ready-dates of the new part numbers or when they become available. "This is not a set-and-forget process," says Maria. "We need to monitor any changes, anticipate and communicate potential issues, and take action."

In the table of the film for the barbecue chips, there is a row with the scheduled receipts. The scheduled receipts are the quantities that the team can expect to receive from suppliers. "There are some formulas that help to determine the optimal quantity by minimizing ordering and holding costs when buying from suppliers," explains Maria.

Economic Order Quantity

"With the example of the film for the barbecue chips, we can see that we have scheduled receipts," says Maria. "While there may be minimum order quantity (MOQ) and order multiple requirements to meet, we want to know the quantity size that maximizes profitability."

The economic order quantity (EOQ)[42] allows companies to identify the ideal size order considering the impact on profits by minimizing the order and holding costs. "Big callout," Maria starts saying, "the EOQ is helpful when the demand is kind of constant, without big jumps, twists, or other acrobatics. If you have fast-growing products, the EOQ application results in…"

"…material shortages," interjects Mike. "We tried EOQ in the past, even for relatively stable products, but it didn't work for us. I'm telling you, I've seen it all."

"Gotcha, Mike. Would you say that it didn't work out because of the data and manual calculations?"

"I'd say that the data we had was terrible, so then we had more and more material shortages. This was like 7 or 8 years ago."

"Thank you so much for sharing your experience. Inaccurate data is a big bump or huge hole in the road for the EOQ implementation that, again, applies to products with constant demand, nothing crazy," clarifies Maria one more time.

When companies enjoy stable demand and data is reliable, the EOQ calculations provide important cost savings and other process improvements. By minimizing inventory holding costs, companies can free up funds to use in other areas to increase revenues and profits. It can represent a big opportunity cost to have money stuck in inventory. Other benefits include better fulfillment, less waste, and less excess inventory.

"Sounds good," says Tim, "but if I remember correctly from class, there's something about quantity discounts that the initial model didn't consider or something like that."

"Buddy, I think you skipped classes at college," says Ruben, smiling.

"To be brutally honest, I did—a few, but I did. Now I'm giving you more stuff to gossip about, right, Ruben?" jokes Tim.

"All good, buddy," says Ruben, and asks Maria, "Is there anything about quantity discounts with the EOQ?"

"Yes, there is. A company can adapt the model to consider quantity discounts, backorders, quality issues, etc. We can use the EOQ even with seasonality; it is more challenging, but it can be done. Oh my gosh, I'm sounding like a politician," says Maria.

"So like a politician too, give us the magic formula that will eliminate all our misery," says Ruben.

"This is the EOQ formula that provides the optimal annual number of orders and the optimal order size when the demand is constant," says Maria, stressing that the demand should be constant while writing down the formula on the whiteboard.

$$EOQ = \sqrt{(2DS/H)}$$

Where
D = demand in units. It needs to be the annual demand.
S = order cost
H = holding cost per unit and per year. Other EOQ formulas include the holding cost as a percentage.

"Mike, I'm sure you agree with me. The real challenge stems from getting the right data, not so much for the calculations. It's true that we have a square root in the formula, but still, it is not scary like when a magician's assistant needs to stay still while getting knives thrown all around. Let's get into defining and quantifying the right safety stock to prevent material shortages," suggests Maria.

Safety Stock

From the operations perspective, there is cycle stock and safety stock. Cycle stock, as its name denotes, consists of the inventory that a company buys and transforms, using the SCOR terminology. Safety stock is a buffer or a cushion for the unexpected, like saving money for rainy days.

Having the right safety stock helps a company combat against sparking spikes on the demand, supplier delays, and forecast inaccuracies, but, if this cushion becomes a whole new large couch with multiple sets of pillows, the company can have too much safety stock. This means greater potential for inventory aging or obsolescence, too much cash tied up, etc. This is not fun. In contrast, too little safety stock defeats the purpose of keeping a buffer.

"There are some formulas to calculate safety stock. You don't need to crunch the numbers for every product; the systems do this for you. But, I think it's helpful to understand the logic behind it," says Maria.

"With all the excess inventory that we have for some potato and tortilla chips, we should forget about keeping safety stock for them; we are sinking in product," says Jon.

"I'm aligned with you, Jon, that in the current situation, for some chips we have too much inventory, but, setting safety stock at zero could cause issues in the future, getting us into shortages. We don't want to go there, either," advises Maria, and adds, "As a precaution, I would suggest defining the right safety stocks. What do you think?"

"True, we may forget to set it back onto the right amount because we are always scrambling with millions of things to keep the operations running," replies Jon.

There are different formulas to calculate safety stock.[43] Choosing the right formula depends, among other factors, on the current and forecast demand, supplier lead times, and inventory velocity. Some of the safety stock formulas are as follows:

- *Basic*—The basic formula of safety stock implies defining how many days or weeks the company wants to cover; for example, two weeks. The calculation is the same as that for inventory days of supply included in the SCOR model[44] multiplied by 15 days.

$$\text{Safety stock} = (\text{Sales in units}/30) \times 15$$

The basic formula doesn't consider demand and lead time, but it's a good approximation.

- *Standard deviation*—This formula applies when several variables are uncertain:

$$\text{Safety stock} = Z \times \sigma LT \times D_{avg}$$

where
Z = number of orders to be fulfilled in a period of time
σLT = standard deviation of the lead time. "We want to know the lead times of each order in the period of time we are considering to then calculate the mean, variance, and standard deviation. We can use Excel for this or a calculator on the internet," says Maria.

D_{avg} = This is the average demand in a period, for example the average daily demand.

- *Variable demand formula*—The variable demand formula works best when the lead time is stable, but the demand jumps around, like when the American singer Pink does aerials and acrobatics in her concerts. The formula is as follows:

Safety stock = Standard deviation of the demand X the square root of the average delay

For the standard deviation part, Excel or online calculators are friends. Calculating the average delay consists of going over the orders in a certain period of time and selecting those delivered late, adding up the delay, and dividing this sum by the number of orders delivered late to apply the square root.

"It's got to be a root or some sort of tree part," says Ruben.

"Of course!" Maria smiles. "The next formula is helpful with rebel lead times because you don't know how they are you going to behave."

- *Variable lead time*—The formula for safety stock with variable lead time applies when the demand is stable, but the lead time is like a rollercoaster with ups and downs and loops:

Safety stock = Z X average sales X σLT

where
Z = the desired service level
σLT = the lead time deviation; "We see this with the standard deviation formula," indicates Amy, looking at her notes.

"That's right. We've met this guy before," confirms Maria. "We can do some additional calculations to ensure that we've got the safety stock right, and guess what? The EOQ is one of them," says Maria, looking at Tim.

Additional calculations related to safety stock are the reorder point and the inventory position.

- *Reorder point*—The reorder point advises companies when they need to reorder to avoid stockouts.

$$\text{Reorder point} = \text{Average stock consumption over a period of time} \times \text{average lead time} + \text{available safety stock}$$

- *Inventory position*—This formula shows the inventory status. If the result is higher than the reorder point, no action is needed. If it is lower, the company needs to replenish to avoid stockouts.

$$\text{IP} = \text{Inventory on hand} - \text{backorders} + \text{inventory on order}$$

Having the right inventory, at the right time, and at the right location fills financial statements with health and joy. Regular checkups with inventory metrics monitor the situation and send signals for future action.

The Baker's Dozen Metrics For Inventory Management

"We start with the strategy, financial metrics, and map out the supply chain metrics by applying the SCOR framework. Within supply chain, we have the power to impact inventory that is a critical area for CPG," states Maria.

Within inventory management, there are many, many metrics. Below are listed the top 13 metrics:

Inventory Days of Supply—Assets

There are different terms that refer to this metric such as days' cost of sales in inventory and days' sales in inventory. This metric is the inverse of inventory turns.

"You can find it in the SCOR DS model, by selecting Performance, after that Economic, and then Assets. The hierarchy of the inventory days of supply metric is AM.2.2,"[45] says Maria.

"Found it," says Sarah, looking at the ASCM online site. At a lower level, there are:
- inventory days of supply - raw material
- inventory days of supply - work in process (WIP)
- inventory days of supply - finished goods
- percentage of defective inventory
- percentage of excess inventory

Per SCOR, the inventory days of supply calculation is as follows:

Five-point rolling average of gross value of inventory at standard cost/annual cost of goods sold/365 days

The site also provides an example: "If two items a day are sold and 20 items are held in inventory, this represents 10 days' (20/2) worth of sales in inventory."

Cash-to-Cash Cycle Time—Assets
Another important metric within assets is cash-to-cash cycle time. "The inventory days of supply metric is part of the cash-to-cash cycle time formula," indicates Maria. Cash-to-cash cycle time measures the number of days from the investment in raw materials until the company makes money from that investment, getting cash back into the business.

The formula is below, found under performance, economic, and then assets in the SCOR model. Its hierarchy is AM1.1[46]

> **Cash-to-cash cycle time = inventory days of supply + days sales outstanding - days payable outstanding**

"Got it! The first part is inventory days of supply that we know how to calculate. How about days sales outstanding and days payable outstanding? How do we get them?" asks Tim.

"What do you guys think?" Maria asks the team.

"The P&L?" asks Mike, "most of the action happens there."

"What are days sales outstanding?" Maria continues the conversation.

"Is it what we are going to receive? Accounts receivable?" asks Amy.

"Where do we have the company's credits in the financial statements?" Maria asks this follow-up question.

"My bad. I see. We get days sales outstanding from the balance sheet, not from the P&L as I thought…and days payable outstanding too?" asks Mike.

"Yes," answers Sarah tentatively, looking at Maria, as if searching for confirmation or approval.

"Let's back up a little bit," says Maria, like hitting 'pause.' "We get sales outstanding or receivables from the balance sheet, but, to do the calculation in days, we need to know the sales that come from the income statement."

"So was I partially right?" asks Mike. "The same for days payable outstanding, huh? We use both the balance sheet for payables and the P&L for COGS."

"Look at you, Mike! You've become a little maniac with the metrics and the financial stuff," says Ruben.

"This deserves an applause," celebrates Maria. "As our award, one more metric within assets for you and the team," she adds, like the presenter in a Hollywood Oscar award ceremony.

Return on Working Capital—Assets
This metric of return on working capital challenges the common saying "you get out what you put in." If this saying held true, the resulting value of the metric would always be one.

"That's wrong, Maria," interrupts Jon. "The ROI varies; it is not one every time."

"I'm with you, Jon, the ROI is not always one or 100%. That's why the saying is not accurate. You can get out more or less than what you put in, don't you think?" Maria asks this question to Jon.

Jon says nothing but shows he agrees with Maria's statement by nodding.

The return on working capital compares the investment on working capital against the revenue generated. In other words, this metric compares what you put in that is the working capital against what you get out, or revenues.

In the hierarchy of the SCOR framework, this metric is AM.1.3,[47] and its calculation is like this:

Return on working capital = (Supply chain revenue - total supply chain management cost (CO.1.1))[48] / (Inventory + accounts receivable - accounts payable)

"Time for metrics on cost," says Maria.

Inventory Carrying Cost—Cost

The inventory carrying cost metric is in the SCOR model under Performance, Economic, to then get at inventory carrying cost, CO.2.3[49] in the hierarchy. This metric focuses on the cost aspects. "The definition is all costs associated with carrying inventory, including opportunity, shrinkage, insurance and taxes, total obsolescence for raw material, WIP, and finished goods inventory, channel obsolescence, and field service parts obsolescence," reads Maria.

COGS—Cost

COGS—metric also with cost focus—encompasses the cost of raw materials and production, including both direct, like raw materials and labor, and indirect, like production overhead.

COGS (CO.1.2)[50] is a level 1 metric included under performance, economic, cost. Its formula is as follows:

COGS = direct material cost + direct labor + production indirect costs

For each of the components in the equation, there are level 2 metrics:
- CO.2.6 direct material cost[51]
- CO.2.7 direct labor cost[52]
- CO.2.8 indirect cost related to production[53]

Total Obsolescence for Raw Material, WIP, and Finished Goods Inventory—Cost

The obsolescence for raw material, WIP, and finished goods inventory is a level 3 metric in the SCOR DS framework. In the hierarchy, this metric is CO.3.21[54] and it is under performance, economic, CO Level-3 metrics. The corresponding level-1 metric is CO.1.1 total supply chain management costs.[55] "I'm sure Ernesto will like this metric," says Amy. "We've had some inventory that we had to write off. If we had known about it earlier, the story could have been different."

"Yes, we can see that metrics help us to focus our attention and alert us when something doesn't look right. Aged inventory is a concern for CPGs," adds Maria.

Perfect Customer Order Fulfillment—Reliability

"Quick recap of the perfect order metric that we've covered when we mapped out the supply chain metrics by applying the SCOR model," says Maria.

The perfect customer order fulfillment metric—RL.1.1[56]—is in performance and then resilience. It relates to service levels. To qualify as perfect, a customer order needs to meet the seven R's criteria, per the APICS dictionary:
1. right product or service
2. right quantity
3. right condition
4. right place
5. right time based on customer's perspective
6. right customer
7. right cost

In addition to meeting the seven R's, the order needs to have all the proper documentation, like packing list, commercial invoice, bill of lading, etc.

The calculation is the total number of perfect orders divided by total number of orders, and this result is multiplied by 100 to get the percentage.

The SCOR model shows various level 2 metrics:
- RL.2.1[57]—Percentage of orders delivered in full to the customer
- RL.2.2[58]—Delivery performance to original customer commit date
- RL.2.3[59]—Customer order documentation accuracy
- RL.2.4[60]—Customer order in perfect condition

Forecast Accuracy—Reliability

The demand forecast cascades to the supply side for teams to create the production plan after assessing capacity and getting the plan into the calendar with the MPS. "We can note that the many errors that a forecast could have also spread into the supply. Considering this, forecast accuracy is quite relevant," says Maria.

In the SCOR DS model, the forecast accuracy metric is in performance, resilience, and then reliability. Forecast accuracy—(RL.3.47 in the hierarchy)[61]—is within the level three metrics. Teams can calculate forecast accuracy at the product or product group level.

$$\% \text{ error} = [(\text{sum actuals} - \text{sum variance}) / \text{sum actuals}] \times 100$$

Inventory Accuracy—Reliability

"Can we trust our ERP or WMS (warehouse management system)? The inventory accuracy metric attempts to answer this question," explains Maria. In performing cycle counts, teams compare the results of the physical counts against the information in the system(s); those that match are accurate."

The calculation is below.

$$\text{Inventory accuracy} = (\text{Number of units on record in the system}/\text{number of units counted}) / 100$$

As a ballpark, inventory accuracy should be over 97% to say that it's at a good level. Inventory accuracy implies a match in units, lot or batch number, and the location. The criteria are stricter than those in finance where the match is in units only, so it doesn't extend to lot numbers and locations. For example, if the system indicates a total of 5000 units in the main warehouse and physically the team has 4500 units in the main warehouse and 500 in a display, for finance, the inventory accuracy is 100% while for supply chain it is 90% [(4500/5000)X100] because there is a mismatch with the location.

Location accuracy is a metric in SCOR under performance, resilience, reliability. It is a level-3 metric, RL.3.28[62].

"Amazing, team! This is our third metric in reliability. If you are ready, we move on to the metrics in responsiveness," says Maria.

Customer Order Fulfillment Cycle Time—Responsiveness

The customer order fulfillment cycle is a full cycle, like a washing machine doing a load of dirty clothes with tough stains. Like the full cycle of such a washing machine, the customer order fulfillment cycle is the sum of cycles. The calculation of the customer order fulfillment cycle is the following:

Customer order fulfillment cycle = sum of actual cycle times for all orders delivered/ total number of orders delivered

As this is a level 1-metric in SCOR, RS.1.1 customer order fulfillment cycle time,[63] there are also level 2 and level 3 metrics in the hierarchy. Each level-2 metric corresponds to a particular cycle time included in the customer order fulfillment cycle.

Level-2 metrics are as follows:
- RS.2.1[64]—Order cycle time
- RS.2.2[65]—Source cycle time
- RS.2.3[66]—Transform cycle time
- RS.2.4[67]—Fulfill cycle time
- RS.2.5[68]—Return cycle time

"A couple of clarifications," Maria mentions:
1. *Non-value added activities included in the calculations*—"In our lean six sigma toolkit we have value stream mapping, where we indicate

value-added or non-value-added activities. In the cycle time computations, we consider them all, no matter that the activity adds zero value."
2. *Dwell time may not be included*—This is the lead time that the customer adds by placing the order in advance, but nothing happens, no activity during this period. "For example, Marcia Williams, the author of this book, places an order for marketing services four months in advance to get on the marketing company's schedule well ahead of time and to facilitate planning. But, the marketing activities (fulfillment) start in the future, three weeks before the book launch," explains Maria. For this reason, there are companies that do not include dwell time in the calculations below.

> **Customer order fulfillment cycle time = customer order fulfillment process time + order fulfillment dwell time (RS.3.97)**[69]

As stated, some companies do not consider order fulfillment dwell time because it is not in their control and doesn't represent inefficiencies in supply chain. In this case, the customer order fulfillment cycle time equals the customer order fulfillment process time. "Out of the five level-2 metrics for order fulfillment, we will dissect (this brings memories of biology classes) two of them: source cycle time and fulfill cycle time," says Maria.

Source Cycle Time—Responsiveness

This metric is a level-2, RS.2.2 source cycle time,[70] and the calculation is as follows:

> **Source cycle time = Authorize supplier payment cycle time + identify sources of supply cycle time + receive product cycle time + schedule product deliveries cycle + select supplier and negotiate cycle time + transfer product cycle time + verify product cycle time**

As Alex's Snacks follows the MTS and MTO strategies, the metrics of identify sources of supply cycle time (RS.3.7)[71] and select supplier and negotiate cycle time (RS.3.10)[72] don't apply to the calculation.

Fulfill Cycle Time—Responsiveness
This is another level-2 metric, RS.2.4 fulfill cycle time,[73] and the formula by SCOR is below.

$$\text{Fulfill cycle time} = \text{MAX}\{[\text{reserve resources and determine fulfill date cycle time} + \text{build loads cycle time} + \text{route shipments cycle time} + \text{select carriers and rate shipments cycle time}] + \text{receive product from source or transform cycle time}\} + \text{pick product cycle time} + \text{pack product cycle time} + \text{load vehicle and generate shipping documents cycle time} + \text{ship product cycle time} + (\text{receive and verify product by customer cycle time}) + (\text{install product cycle time})$$

If companies adopt MTS or MTO like Alex's Snacks, per SCOR, the optional metric of schedule installation cycle time (RS.3.27)[74] doesn't apply.

% of Renewable Materials Used—Sustainability
The revision to the SCOR model in the last quarter of 2022 includes the sustainability area. One of the level-1 metrics under environmental is EV.1.1 materials used.[75] This metric refers to the total weight or volume of materials that are used to produce and package the organization's primary products or services, per the SCOR's website.

This metric has a couple of metrics in level 2 in the hierarchy. They are:
- EV.2.1[76]—renewable materials used
- EV.2.2[77]—non-renewable materials used

"And now, my dear team, another Chips in Action," introduces Maria.

Chips in Action

1. Define your current levels for each of these 10 metrics.
2. Benchmark against history.
3. Benchmark against your industry.
4. Identify any gaps.
5. Follow the approach of lean six sigma and apply the tools from your toolset to identify the root causes and solve the issues.
6. Measure the new levels for each of the metrics.
7. Assess what's working and what's not working.

Procurement and Procure to Pay, extending into Logistics

Being on the supply side of the equation, procurement has an umbilical cord with finance and the financial statements. "We have been touching on procurement and cost savings here and there because of its extended impact. We are now all hands-on deck on procurement, a function with great potential for increasing the business's profitability," says Maria.

There is confusion and misunderstanding about the supply chain term, but procurement doesn't seem to have the same issue. The job of someone in procurement is to buy amidst the multiple complexities involved. They are shoppers, as Elise, Maria's middle daughter, refers to the procurement team members. This means that they open the company's wallet to buy products and services, taking away the precious funds that the finance department treasures.

Because of the nature of the relationship between finance and procurement, there are some discussions. Finance may question how procurement is using the company's funds and if procurement can get the items and services at lower prices through discounts or buying somewhere else; it's pretty much like a husband and wife arguing about holiday shopping.

Just as spouses can have a harmonious relationship, finance and procurement can and should be happy together. For this to happen, the name of the game is ROI.

When one spouse wants to convince the other one, she sells him on the benefits, and these benefits, of course, should be also benefits for the other spouse. For example, one of them wants to go to Rome on vacation and the other one to Machu Pichu in Peru. To make the case for Machu Pichu, the spouse can say that they can spend more time together there enjoying nature, as on the trails the internet connection is spotty at best.

Procurement sells finance on the benefits, on the ROI, showing the impact on the metrics included in the strategic goal tree that contains the financial and supply chain metrics aligned. After working hard on projects, sourcing managers sometimes provide finance with too many details related to operations and don't include a snowflake crystal ROI calculation. This makes the savings melt before having any chance to solidify in the financial statements.

"The example below shows different approaches. Which one would you choose?" asks Maria.

"Aha! This is new, like a multiple-choice question," says Ruben.

"It is!" confirms Maria.

Sourcing manager says to finance:
Approach 1: This is a critical project because before we were buying the corrugated boxes from supplier X, which was getting the raw materials from Z. Z increased prices to keep the same lead time, as they were struggling with capacity. X increased their prices, and the increase was quite significant. We went to the market and asked for a quote.

Approach 2: This project is important to me because if I could get 12% in savings, I would meet my target and who knows, maybe I could get a promotion from Ernesto.
Approach 3: We estimate the ROI of this project is 22%, and here's why (showing a summary).

As the number three approach is the obvious correct choice, Maria highlights some key points of each approach. With the first approach, finance gets bored with details that are not relevant. In the second approach, it is all about the sourcing manager. This is like selling the trip to Machu Pichu saying because that's what I like and that's what I want. The chances for success are low.

The right approach, number three, goes directly to the ROI, speaking the same language as the business and finance people, and shows a summary before getting into too many details or spreadsheets.

"The main hurdles for procurement are identifying impactful cost savings opportunities, quantifying them in the right way, and executing them. We are going to address this, team," says Maria. "We have key areas including:
- Total cost of ownership,
- Effective ways to achieve cost savings,
- Estimate cost savings in challenging situations and how to present them to Finance,
- RFPs and legal language to include to be protected, and
- Opportunities in Procure to Pay."

Total Cost of Ownership

"A picture is worth a thousand words. We want to over-deliver but not under-promise, as Grant Cardone indicates in one of his best-selling books, The 10X Rule. Let's then have two pictures instead of one," says Maria.

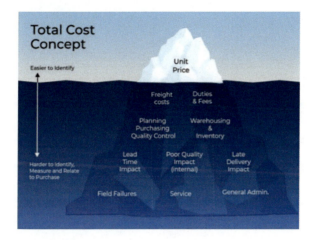

Source: Michigan State University

Both illustrations show that only a small part of the costs are visible, the top of the pyramid or the tip of the iceberg. You have to delve into the intricate pyramid with the determination of Indiana Jones to see the other costs. The picture with the iceberg shows the same concept; the naked eye can see about 20%.

The total cost of ownership (TCO) concept consists of adopting a holistic approach to consider 100% of the costs, reaching the 80% of costs that are under water or at the bottom of the pyramid. "I know, I know," says Maria, looking at Jon lowering his eyebrows and tightening his lips, "it's challenging to estimate costs like lead time or poor-quality impact, but we have our toolset. Estimates are not perfect. It's better to have estimates that reflect our thought processes than not considering those factors at all."

"How do we do this TCO? I know about the nice toolkit we have, but at this moment, I feel like I am in the middle of an aisle at Home Depot, with hundreds of tools around me from Stanley, Black & Decker, and Bosch, and I don't know if I need to pick a hammer or a drill," says Jon.

"Hi, how can I help you? I'm Maria and will be happy to assist you at Home Depot. A tool that can help you with TCO is the Finished Goods Margin Simulator. With a Chips in Action, we'll put this into practice," answers Maria.

Chips in Action

1. Get your Finished Goods Margin Simulator from the resources page at https://usmsupplychain.com/resources/.
2. Identify other costs to include, if any, and add them to the simulator. Hint: when you don't know what to do, don't despair, follow the DMAIC approach and use the tools in your toolkit.
3. Work with finance and manufacturing or operations to define the parameters to use and get your initial TCO model. "Bear in mind that estimates are not perfect, but they are better than nothing. You are doing this, team!" Maria says, switching her role from Home Depot sales assistant to cheerleader.
4. Implement your TCO model and make adjustments, like rinse and repeat.
5. Enjoy the well-deserved flavored potato chips!

While the team is taking a break and engaged in casual conversation, the door opens as if it's being pushed by a strong tornado. It's Ernesto, hopping from meeting to meeting all day, and in a hurry.

"I think we have too many meetings. I believe it's the culture in corporate America. Not sure if this is the most productive way, but that conversation is probably for another day. I want to focus on this point of the agenda, fifteen effective ways to achieve cost savings," says Ernesto, settling down in the conference room and drinking some water.

"Ernesto, based on your comment about meetings and busyness, I'd like to recommend you a book. It's called Deep Work by Cal Newport. It's totally worth it. Have a read and then please let me know your thoughts," says Maria. The team proceeds with the training.

The Baker's Dozen Effective Ways to Achieve Cost Savings— Examples with Logistics and Fulfillment Included!

Without further ado, pulling back the curtain, here they are, the fifteen effective ways to achieve cost savings.

Ask suppliers for cost reduction.

"Is this really a method?" asks Amy.

"Yes, it's a simple and powerful method that companies may miss out on," answers Maria.

This method works well in two distinct situations:
1. Changes outside of what companies can control, like recessions, new regulations, pandemics, and
2. If it has been some time since the last time the company ran an RFP or negotiated with the supplier.

"It works. Plain and simple," says Maria, who has witnessed the results of asking suppliers for cost reductions during the Great Recession[78] from December 2007 until June 2009 and with the tariff changes to China in 2018.[79]

"How is that possible that suppliers reduce prices in such gloomy conditions?" asks Mike.

"They do, Mike. Precisely, in the communications with the suppliers we highlight the unfavorable conditions. The words and the narrative are critical. You would be surprised how suppliers respond," states Maria, and adds, "Ernesto, we did this with The Walls and the results were off the charts!"

"Yeah, Tom was pleased with the outcome. He told me that they got cost reduction from suppliers they hadn't talked to for years," confirms Ernesto.

"So glad to hear that. Yes, with David and Tom we were in the second scenario," explains Maria.

"Would it be possible to do something like that here at Alex's Snacks?" asks Ernesto.

"Of course, getting into it right now with Chips in Action. Let's get some savings, team," says Maria.

Chips in Action

1. Obtain the spend by supplier for products and services—direct and indirect spend. "To get the spend, we sometimes need to play Sherlock Holmes, but there is no mission impossible, guys, with this step-by-step process," Maria explains.

> How to Obtain the Annual Spend of Products and Services Step-by-Step and What to Watch out for:

- Get the list of valid POs from the last 12 months. Valid means that they are POs with receipts against them, or that they are open because the supplier hasn't shipped all or still hasn't shipped. The idea is not considering those POs that are open, but we don't actually expect any receipts.

- Some companies use applications like Ariba for indirect spend. Ariba can categorize the spend in different buckets through commodity codes. "It's important that the team selects the right commodity code when creating purchase requisitions for Ariba to work like a clockwork," says Maria.

- Although not advisable, some companies may not use POs for services. They use invoices only. In that case, the team can review the invoices. The preference would be to obtain the spend data from the POs, as they have more information than the invoices. But, if there are no PO's, like with services, the team can get the data from the invoices.

- When companies use blanket POs, the spend details are not in SAP or the ERP that the company has. The team needs to get this information from a supplier's report.

- A similar situation takes place with credit card payments. The details of the spend are not in SAP either. Like with blanket POs, the team needs to get the spend data from the credit card statements.

- Along these lines, some POs may not have line item detail. In this situation the team needs to ask the supplier for the spend information with more granularity.

- Like other CPGs, Alex's Snacks works with co-manufacturers and co-packers. If the co-manufacturers own the inventory, this inventory is in their systems. In this case, the team needs to ask the co-manufacturers for the spend data.

"With macros in Excel, we can get all the spend data together from the different sources with minimal manual input to ensure accuracy—and get it done fast," explains Maria.

"We receive reports in PDF format; will the macro in Excel work?" asks Sarah.

"Yeah, we have suppliers that send the details of the blanket POs in PDF. The same with some co-manufacturers," confirms Amy.

"We can convert those PDF files into Excel. We can use Power BI or there are several online converters like https://pdftables.com/[80] or https://www.ilovepdf.com/,"[81] indicates Maria.

- There may be issues with the part numbers too. For example, the same product or material could have two or more different part numbers. This could happen because the team uses supplier part numbers or because there have been inconsistencies in the naming convention. Putting the spend together is more challenging in this case. "We want to clean up the item master as much as possible by assessing benefits versus effort," Maria says.

- The company may have the same issue with supplier names when compiling the spend data by supplier. "We can have PChips, P-Chips, PChips Inc., pchips, etc. All these names for the same company. Can you relate to this?" Maria asks.

- "You can't imagine how," says Amy, shaking her head.

- If there is no trust in the company's systems or they don't have the required spend details, building a simple template in Excel for the suppliers to complete at the end of each year can be a good solution. The company's spend data is the suppliers' sales data. "I'd say that the suppliers know their sales, don't you think?" Maria asks the team. She has seen this approach in Fortune 100 companies. The team can also use the supplier's data to validate accuracy.

2. Rank the spend by supplier, starting with the highest spend.

3. Include the products the company is buying from each of the suppliers, quantities, cost, and details about the last supplier negotiations.

4. Select the 20% of suppliers that represent 80% of the spend. This is wave one, the suppliers to focus on. Create other waves with the remaining suppliers.

5. Compose a powerful personalized email to request cost reduction to send to wave one. "This email needs to show that you know the specifics of the relationship with your supplier. It is not a chip-cutter approach."

6. Track your progress in a template.

7. Quantify estimated annual savings showing total and timing. "More about timing when we get into how to present the savings to finance," clarifies Maria.

"We need to do this immediately, team. If you need any support, please come to Maria or me to make it happen. It is a top priority for us to deliver cost reduction," says Ernesto, standing up and getting into the front of the room in the blink of an eye.

The team works behind closed doors to get the spend details and get this approach into motion. Maria has more effective ways for the team to get savings for the following days.

Better Planning

Another effective way to obtain savings is with better planning. Like Dwight D. Eisenhower, World War II leader and 34th president of the United States, said, "In preparing for battle I have always found that plans are useless, but planning is indispensable."[82]

Plans are wrong, like forecasts. The longer the horizon, the worse it becomes. But, with the planning process, companies develop a thorough understanding of options and constraints. This makes them better prepared for decision-making when uncertain future events unfold.

A clear example of where companies can get good savings through better planning is logistics. "The stakeholders sometimes want the moon, the stars, and a comet or a couple of them, like having a shipment overnight from China," says Mike, and adds, "There are direct flights from Shanghai to LA or to New York, but how about customs? It's not like stuffing a plane like you do a Thanksgiving turkey, and when the plane arrives, they deliver the cargo right to you. I wish it could be that easy, but it is not, people."

Mike's frustration comes from a misunderstanding of the processes and lead times of the stakeholders. Logistics should be simple, going from point A to point B, but there are many more factors than the ship from and ship to locations.

"We need to map out the processes, and we have the tools to do so," says Tim.

"Excellent, Tim! Yes, we do have processes, plural, as there are many complexities involved," Maria answers.

In logistics, there are inbound and outbound shipments. Inbound are those shipments with raw materials getting into the business. For example, when the company buys film for packaging, these are inbound shipments. Outbound are those shipments that go to customers like Walmart, Costco, Albertsons, Target, etc. When selling direct to consumers, B2C, these shipments are also outbound.

Another distinction is between small parcel shipments and regular palletized shipments like shipping LTL (less-than-a truckload) or FTL (full truckload). The difference is in the shipment size. "We know FedEx, UPS and USPS because of small parcel shipments. We have Amazon Logistics, too, and there are other smaller companies," says Maria.

"We now turn this into a logistics class—an unexpected turn that to the best of my knowledge is...stupid, because Mike knows all of this and these differences. What are you trying to do here?" asks Jon, lowering his eyebrows and tilting his head downwards.

"Jon, we know that Mike is great in what he does. No question about it. Maria is explaining the basics or fundamentals of logistics so all of us can be aligned," says Ernesto.

"Thank you, Ernesto," replies Maria. "Jon, my intent is not to teach logistics. I am mentioning these points to help map out the different processes and ask the stakeholders the right questions to avoid frustrations and misunderstandings. Good outcomes follow good inputs," Maria says and continues with the modes of shipment.

The main modes of shipment are air, ocean, and ground, which includes rail and over-the-road (FTL, LTL) with different characteristics, for example, speed and cost. Shipments can be domestic or international. International shipments are complex because of the regulations and requirements that they entail. "This means that in planning, we need to consider different scenarios, lead times that are longer than those in domestic shipments, and the required documentation," clarifies Maria.

Whether domestic or international, shipments have three main flows:
1. *Physical flow*—This is the cargo, the products that move.
2. *Cash or money flow*—"We need to keep track of the true cost to identify and execute on opportunities," says Maria.
3. *Shipping documents*—"That's a big one for us," says Mike.

"We get shipments stuck at customs after paying for expedited shipping because something is missing. We sometimes don't know the reason for the clearance delay. All the rush for nothing" says Tim.

"And we get quite expensive bills to pay because of expedited shipping," says Amy.

"Great! I mean, great to know that we have an opportunity here, a potential quick win if we define the requirements of the shipping documents," says Maria.

Considering the team's inputs, Maria suggests the following action items for better planning of the inbound shipments:

- Map out the inbound shipments, noting the distinction between domestic and international, small parcel or palletized, and mode of shipment.
- For each step, show the average lead time in days.
- List the different options available, indicating definition, lead time, and estimated cost. For example, FedEx has various options for overnight:
    - overnight standard—FedEx can deliver the shipment at EOD (end of day). "This means that the team can start using the materials the following day, not on the delivery date," clarifies Maria.
    - overnight am—The arrival time of the shipment is during the morning.
    - overnight first—The shipment arrives first thing in the morning, at 8 am.

A similar example is with international shipments: international economy (one to two weeks) and international priority (three to four days).

"We want the definitions to be specific for better planning and to avoid any disappointments, like when a girl is waiting for an O.M.G doll but she receives pajamas for her birthday present," says Maria.

Lead times and definitions are crucial. Imagine the plant expecting to receive the materials on Monday morning but getting them at the end of the day. Big difference. Big disappointment.

- Model different scenarios indicating lead time and cost.
- Create thorough standard operating procedures (SOPs), including the requirements for the shipping documents and the location of the different files. "We can apply 5S, team, and ensure we keep the proper documentation in place, easily accessible when we need it," suggests Maria. "We have so many files, maybe, too many," says Sarah and adds, "it can take quite a bit of time to locate a file."
- Build a library with FAQs. This facilitates onboarding by shortening the learning curve. For example, it is important to include tips about past shipments, including the lessons learned. For example, splitting shipments with the essential quantity coming via air while the bulk of the shipment comes via ocean.

"You can follow a similar approach for outbound shipments, guys," says Maria.

"Do we take this as the next Chips in Action?" asks Ruben.

"You bet. We need savings, and by improving planning, we have great opportunities," says Ernesto.

Chips in Action

- Follow step by step the actions suggested above for better planning of inbound shipments.
- Create a similar approach for outbound shipments.
- Do the same for any other supply chain and procurement planning process after building the priority matrix you have in your toolset.
- Estimate annual savings and timing.

"What else have you got, Maria, to save us money?" asks Ernesto.

"The next effective way to achieve cost reduction is supply chain network optimization," says Maria.

"I don't know about you guys…my gut is telling me that formulas and numbers and trees are coming back. Well, actually, they have been here with us all the time," jokes Ruben.

"All right, let's do it!" says Maria.

Supply Chain Network Optimization

Supply chain network design[83] or optimization consists of building a model with a certain logic that replicates the supply chain with actual transactions and transformations. By applying the model, companies can assess different aspects of the supply chain and identify opportunities for process optimization and cost reduction.

"There you go! Math, data, cost savings, what else can we ask for? Maybe a hidden tree?" says Ruben.

"Agreed, Ruben! We are having a blast!" says Maria with a smile.

Supply chain optimization is a structured, holistic process that uses mathematical and analytical models to determine the best combination of facilities, suppliers, and products. Like with any other model, there are inputs and outputs.

The company needs to define its business objectives including its markets, its growth plans, and its financial objectives. There are granular inputs such as customer service levels and pricing. Like the wizard Merlin with his wand, the model works its magic to provide the best combination of supply, production, and distribution costs.

Doing this modeling in spreadsheets is cumbersome and time-consuming because teams need to adopt a holistic approach. A holistic approach is unattainable when the different functions work in silos. Companies can have a WMS (warehouse management system), TMS (transportation management system), IMS (inventory management system), CRM (customer relationship management), and other software suites, including manufacturing and procurement. But, the silo approach doesn't lead to an optimized solution for the end-to-end supply chain.

Like ninjas breaking boards and bricks, S&OP demolishes the functional barriers. "We know that for S&OP to be effective we need to glue it with finance, with loads of Krazy glue and Super Gorilla. We want IBP, the most evolved version of S&OP that focuses on the financial impact. Raise your hand if you agree," says Maria.

While all the team members are raising their hands, Maria asks "If you are anything like me, you want to understand how the magician does the trick behind the large black box. Are you interested in finding out how the model gives the optimal combination of facilities, suppliers, and products?"

"I don't know. Sometimes it's better not to know," says Tim.

"I'm in. I want to know," says Ernesto.

"Kind of expected from you, Ernesto. And you ladies?" asks Ruben.

"Yeah, we want to know," answers Amy in agreement with Sarah, who is nodding her head.

"The trick is prescriptive analytics," reveals Maria.

Prescriptive analytics is within advanced analytics. Advanced analytics consists of the following:

- *Descriptive*—What happened? Per the Gartner Analytics Continuum, descriptive analytics is the most basic form of advanced analytics, like the ape in Darwin's theory of evolution. It provides answers to questions like how much profit the company made, last month's sales, the company's largest customers, how many cases of barbecue chips are sold, the percentage of Amazon sales—you get the idea," says Maria.

- Descriptive analytics illustrate what happened in the business, as they use business intelligence and data on past activity. "Examples are the metrics we track," clarifies Maria.

- *Diagnostic*—Why did it happen? This is the question that the Analytics Continuum shows as the next step in maturity. The company knows the numbers with descriptive analytics and now wants to understand why those numbers, the reasons behind them. "Like when we follow the SCOR DS model to have our strategic goal tree and identify what the metrics tell us. We then pull the root cause tools—diagnostic analytics—from our lean six sigma toolset to dig deep," says Maria.

  "I knew it! The trees! We can't make improvements without them," says Ruben.

"We can't make improvements; we can't live without them," replies Maria.

Both descriptive and diagnostic analytics are based on the past. As Mike has shared, Alex's Snacks experiences delays with its inbound shipments. The number or percentage or delays in number of days is descriptive analytics. One of the reasons is errors in the shipping documents. This is diagnostics analytics. Both the delayed inbound shipments and the errors in the shipping documents come from past shipments.

"Questions?" Maria asks.

"We are good," says Ernesto.

- *Predictive*—Per the Gartner's Analytics Continuum, "What will happen?" is the representative question of predictive analytics. With predictive analytics, companies use forecasting techniques and statistical models to advise what could happen. Predictive analytics anticipate future events so the team can be prepared and take action.

  Continuing with the inbound shipment example, the team builds a model that predicts the ETA (estimated time of arrival) considering different factors. This model is predictive analytics.

  "You have tools in your toolset with the forecasting techniques that use predictive analytics. I want an example, team," says Maria.

  "Regression analysis? I remember that it is machine learning and AI," says Sarah.

  "You are a sophisticated lady with the AI terminology," says Ruben.

  "Outstanding, Sarah! Yes, regression analysis is predictive analytics that consists of algorithms, what-if scenarios," says Maria.

- *Prescriptive*—How can we make it happen? This question shows the highest maturity level in the Gartner Analytics Continuum for prescriptive analytics. Predictive analytics anticipates issues, challenges, anomalies in the supply chain network. Prescriptive analytics takes it from there, from predictive analytics.

There are two types of algorithms in prescriptive analytics: 1. heuristics (rules) and 2. exact (optimization). Heuristics algorithms are shortcuts to get good answers in a reasonable amount of time. Heuristics don't always provide the best answer, at least, there is no mathematical way to prove that. "They are helpful in situations where we can apply rules of thumb; if this happens, do this," explains Maria. Examples are buying more raw materials when the stock level gets to this X quantity, or assigning capacity first for the original, second for cheese curls, and then for the barbecue chips.

Picking up the logistics example again, if the model predicts a traffic jam or bumper-to-bumper traffic because of planned road maintenance or construction work, prescriptive analytics reschedules the route for the driver. "Each kind of analytics builds on top of the other," says Maria.

Exact algorithms or optimizations like linear programming, in contrast, provide the best answer, as they are based on a proven scientific technique. With these exact or optimization algorithms, prescriptive analytics indicate what should be done, not merely what is possible.

There are many good feasible scenarios, but prescriptive analytics with the optimization algorithms provide the best scenario through embedded decision logic rules, considering business objectives, variables, and inputs.

Getting to the optimal solution requires time. Examples are minimizing the logistics cost of inbound shipments or what the product mix should be.

"No Chips in Action on this, as it would require software and entering all the inputs, right?" asks Amy.

"About the software, we can use solver in Excel for prescriptive analytics. Happy to help the team," answers Maria.

The team and Maria move to the fourth way of achieving effective savings: right sizes in buying, manufacturing, and shipping.

Right Sizes in Buying, Manufacturing, Shipping

Right sizes in buying, manufacturing, and shipping are part of the supply chain network optimization. "After the overview and revealing how the magic trick is done, we go deeper into buying, manufacturing, and shipping," says Maria. The approach should be holistic, considering supply chain networks, as opposed to an in-siloes approach.

In isolation, with a constant demand, a regional warehouse can apply the economic order quantity formula to minimize holding and ordering costs when requesting products from the central warehouse. The decisions of that particular warehouse without considering what's available in other warehouses, the different storage costs and capacities, may not be the optimal decision.

The concept of multi-echelon (multiple tiers or layers) inventory optimization considers the supply chain as a whole, taking into account, among others, storage cost and capacity limits of the entire supply chain network to arrive at the optimal solution.

In manufacturing, there are three main production systems:
1. *Mass or flow or continuous*—This is a non-stop massive quantity produced, like with the original potato chips at Alex's Snacks. CPGs use mass production for several of their brands. The optimal use of the resources—workers and machines—is fundamental.
2. *Batch*—Batch production is between mass and jobbing. It is for medium quantity and medium variety. Batches apply to MTO (made to order) when companies offer customization. CPGs also produce in batches. "An example is lipsticks with so many colors," says Maria. In batch production, the batch size plays a pivotal role. Having the wrong batch size could cost companies a lot of money.
3. *Jobbing*—Jobbing is for one-of-a-kind products, based on a customer's specifications. These products are seldom repeated.

Although Alex's Snacks doesn't own manufacturing plants, conversations on the manufacturing process and performance can lead to win-win situations.

Shipping is another area with potential savings. "I'm proud of your work in this area, team!" says Maria. By applying optimization tools, for instance, to load containers, trucks, planes, companies can realize significant savings.

The holistic approach applies to new product introductions. When designing a new product launch, the engineering or R&D team sometimes works in a vacuum. This could lead to costly mistakes. For example, an innovative packaging design could have a negative impact in manufacturing and then in shipping the new products.

Another example is with the storage fees when Amazon provides storage services for companies selling online. There is a table named product size tiers.[84] This is one of the factors, together with current month, current volume, average daily units, and dangerous good classification status that Amazon uses to calculate storage fees. If the team doesn't consider the product size when launching a new product, the storage cost can be as difficult to swallow as very spicy chips without a drink.

The team gets ready for another effective way to save cost that is process optimization.

Process Optimization

For fast growing CPGs, it is imperative to optimize their procurement and supply chain processes for healthy, profitable growth. They are like those plants outgrowing their pots with their roots visible. This is when streamlined processes and documented SOPs (standard operating procedures) become a need for survival and "the next level" goes from being a generic marketing term to sell to a reality for supply chain.

Fulfillment operations are a good example. "Think about going from fulfilling hundreds of orders per day to thousands!" says Maria. "The change is brutal. Many manual processes that made the company grow have to go now, as there is no time for that. This is when the plant doesn't fit its pot any longer and calls for structure, frameworks, SOPs."

"Bravo! I was missing our trees, plants, and roots," says Ruben.

"Please add low hanging fruit to your list, Ruben!" replies Maria.

In these situations, there are savings to easily identify and catch. The company may use automation in Excel to manage the growing data sets and build reports in Excel or Power BI to detect these opportunities before investing in expensive software.

Maria says, "The savings are in the process, not in the implementation of new technologies. Making a process faster doesn't make it good."

Tim says, "So what do we do here? Do we use the tools for process redesign and the DMADV?"

Maria says "Gorgeous! The mistake I often see is that to fix the overwhelm that comes with growth, companies pay generous dollars in hope of stopping the pain. Process and profit first. Then technology."

The team is ready for another effective way to achieve savings—in-house vs. outsourcing decisions.

In-house vs. Outsourcing

Decisions on inhouse vs outsourcing or buy vs make apply to different areas within the supply chain. Like Alex's Snacks, in their beginnings many companies outsource manufacturing because there is uncertainty if the future company's sales will be able to afford the manufacturing costs.

When production volumes keep going up, and get to a certain stable volume, teams need to reassess this decision, as in-house manufacturing can provide both lower cost and higher sales. On the sales side, with inhouse manufacturing, companies can respond faster in addressing issues with their existing items or creating new products. The company is in control of the production schedule, having more flexibility.

On the cost side, in-house manufacturing can lead to cost savings, eliminating the middleman. In addition, as the business has more control over processes and costs, in-house manufacturing can lead to savings.

"Are you suggesting that we should leave our co-manufacturers hanging after all their hard work to build this business to add a humongous overhead cost?" Jon asks.

"I'm suggesting an assessment. I understand, Jon, that bringing manufacturing in-house implies property assets, managing labor, benefits, and worker's comp claims. There are pros and cons," Maria replies.

Jon's eyebrows furrowed. "This is working well. Why change it?" he asks.

"It has worked great. But now we are having issues because the company has outgrown systems and processes, like the plant its pot," Maria answers.

"Is the next Chips in Action the assessment on in-house manufacturing vs. outsourced?" Tim asks.

"You're ahead, Tim! Please go ahead. In the Chips in Action, please keep in mind the buy vs. lease for CAPEX (capital expenses), for example, machinery or equipment," Maria replies.

Like this assessment on in-house manufacturing, the team can do a similar evaluation for the logistics area and the potential use of the vendor-managed inventory model, where inventory management shifts to the vendor.

With the completion of the Chips in Action, the team moves on to the next effective way to achieve cost reduction—MBE (Minority Business Enterprise) suppliers.

MBE (Minority Business Enterprise) Suppliers

Having more MBE suppliers is another effective way to achieve cost savings.

"Really? This is a given because you are one of them," says Jon, then pausing for a moment to add, mimicking Maria's voice, "Hire me, hire me, because I am an MBE. I bring savings."

"My suggestion is based on research by The Hackett Group and McKinsey," replies Maria.

In 2017 the Hacket Group published an article[85] addressing the concerns about bringing MBEs on board because of loss in efficiency. Their research indicates that top performers are not seeing such a loss, and the risk of hiring MBEs is low, with a great upside potential. The Hackett Group's research suggests that not having an MBE program can result in a loss of sales.

In its paper, McKinsey shows that MBEs[86] can have a positive impact on the bottom line with the savings they bring. Per McKinsey's research, MBEs bring their corporate partners' year-over-year cost savings to 8.5 percent, more than the 3 to 7 percent annual procurement savings that most organizations realize.

The next effective way to achieve cost savings is related to this one with ongoing supplier evaluation.

"As we are an elite performance team looking for efficiency, I consider that we can combine the Chips in Action for both ways," says Ruben with a serious look on his face.

"You've changed, Ruben," says Sarah.

"Well, it is my formal way to say that we are a kick-ass team," replies Ruben.

"Team, you are rock stars! I'm so proud of you!" says Ernesto.

The team decides to do the Chips in Action combining MBE suppliers with ongoing supplier evaluation.

Supplier Segmentation

With a higher number of purchases to accompany the company's fast growth, the supplier base can get out of hand. Procurement professionals need to take a deep breath and then get their heads around understanding the current situation or baseline. Knowing the supplier database helps companies identify opportunities for cost savings. A framework or matrix to use is the Kraljic Matrix.

The Kraljic Matrix is an effective way for supplier segmentation. Supplier segmentation is the initial step of SRM (supplier relationship management), a fountain of savings when companies design strategies based on this map of suppliers.

Kraljic indicates that teams should map the supply items against two key dimensions: risk and profitability.

Risk refers to the chances for an unexpected event in the supply chain to disrupt operations. The Russian-Ukrainian war and its impact on oil is an example. It has had a tremendous impact on the supply chain of Alex's Snacks.

Profitability relates to the impact of a supply chain item on the company's financials, the language of business. Oil has a high impact on profitability, while the Post-It-Notes the company buys have a low impact. "I think we can live without Post-It-Notes," says Amy.

Here's how the matrix combining these two dimensions looks:

Source: Peter Kraljic, HBR

Each of the quadrants represents a different buyer-supplier relationship that grants a different sourcing strategy:

Non-critical items—These are low-risk, low-impact on profitability. The focus is on efficiency. These items are excellent candidates for reverse auctions. "Bear with me," says Maria, "we will be there in a few minutes."

Leverage items—These are low-risk, high-impact on profitability. Although these items have a high impact on profitability, buyers have an advantage because there are other suppliers from which to buy.

"It's a nice position to be in," says Tim.

"Does it mean that we can squeeze suppliers to get savings?" asks Mike.

"That's the most common approach. I have also seen companies that work with their suppliers to get savings year-over-year with creative solutions," says Maria.

"I see potential in going through the supplier segmentation process with this matrix to identify each type of relationship and get savings. We haven't done so in a while," adds Ernesto.

"Glad that you find this helpful, Ernesto," says Maria, "we have two more pockets to go over."

Bottleneck items—These items are high in risk and low in profitability impact. This is the opposite of leverage items. This means that the supplier has the upper hand.

"I don't like this. Can we skip the bottleneck items?" asks Ruben.

"Mm, I know the feeling, but we can change this position. We need creativity," replies Maria.

Bottleneck items demand time and energy from procurement. Procurement sometimes doesn't have other options than to accept and live with ugly conditions. In these cases, procurement seeks to limit the damage by altering the terms of trade or getting out of the situation with the redesign or redevelopment of the product.

Strategic items—These are high-risk, high-impact on profitability. These are THE items because they are critical for the business. Procurement should center its efforts on managing this handful of suppliers. Both parties need to work together to build a long-term relationship and foster innovation, continuous improvement, and savings (of course!).

Applying this matrix in the right way is critical. Treating non-critical suppliers as strategic ones leads to a waste of resources and the corresponding negative impact on profitability. Likewise, treating strategic suppliers the same way as non-critical suppliers translates into massive lost profits.

"Now with you, this Chips in Action covers two effective ways to save costs: spend with MBE suppliers and supplier segmentation," says Maria.

Chips in Action

1. Categorize all suppliers per Kraljic's matrix.
2. Identify MBE spend.
3. Identify opportunities to increase MBE spend.

Quick Requests for Quote

Quick requests for quote apply when there is not a designated vendor, and the purchase needs to be fast. This is a good approach for the leveraged items in Kraljic's matrix.

With the specifications and quantity order, the company requests quotes from three or four suppliers. If all the suppliers meet the specifications, the buyer selects the company with the lowest price. This approach ensures that the buyer does the due diligence, removing subjectivity in the selection process.

For complex items and services, the company can launch an RFI (request for information), an RFP (request for proposal), and or an RFQ (request for quote).

RFXs (RFI, RFP, RFQ), including Online Auctions

Companies issue RFXs when going to the market for critical items and for complex services. In general, the result is a two-to-three-year contract.

Because the RFXs are thorough documents with the company's confidential information, the company sends an NDA (non-disclosure agreement) for the suppliers interested in participating to sign. "As terminology varies, we start with defining what these acronyms mean," says Maria to introduce this effective way to achieve cost savings.

RFI—It stands for Request for Information. Companies can issue an RFI using electronic tools like Ariba, for instance, or via email. The RFI is exploratory and includes questions whose answers are not on the supplier's website. For example, questions about the business model.

RFP—This is a Request for Proposal. An RFP can follow an RFI, or a company can issue an RFP from the get-go. An RFP is a structured document package that consists of the following:

- *Project overview*—the objective of the project overview is to furnish the suppliers participating in the RFP with a high-level view of the project, including the category, context information, and timeline.

- *Legal contract sample*—This is the boilerplate or standardized language that the company uses. This is helpful to shorten the contract negotiations, as the suppliers are aware of the expectations, and the company can know any sticky points when receiving the suppliers' proposals.

- *Business requirements*—Business requirements are the "must-have's." Failing to meet any of the requirements means the supplier is eliminated, like in the reality shows—out of the house, off the island, or out of the race to wed. Business requirements are critically important because they become part of the contract once the selection process finalizes.

- *Information questions*—As the suppliers should meet all the business requirements, the information questions allow the company issuing the RFP to differentiate suppliers' capabilities.

"Asking the right questions is fundamental," says Maria. In creating the questions, we need to think about what we expect to get and how we are going to assess the supplier's answers. For example, is it important to know the suppliers' exact sales to similar companies, or would it be more meaningful to have ranges for the suppliers to select?

Defining the objective and anticipating suppliers' answers are important in open-ended questions. As suppliers' answers can fill a book, procurement needs to formulate the question in an appropriate way to get only the concise information needed.

Companies can create a library with requirements and information questions to use in future RFXs. This allows for a quicker preparation of the RFXs documents and leverages the knowledge gained with lessons learned.

- Pricing section—The pricing section contains a pricing worksheet for all the suppliers to fill in. Having the suppliers complcte the pricing worksheet with the same format facilitates the compilation of suppliers' answers and an apples-to-apples comparison.
  In addition to the pricing worksheet, the company can allow suppliers to provide an alternative pricing proposal if they consider it beneficial. With the pricing worksheet and the alternative proposal, the company covers its basis to have an accurate comparison with all the proposals and assess alternative proposals too.

RFQ—It stands for Request for Quote. It may include an online auction. Online auctions have been around since the mid-1990s.

Like with Rocky Balboa, the secret of a successful online event lies in the preparation. For example, before the big day, procurement needs to make sure that all suppliers submit pre-bids that are equal to or lower than current prices. "We don't want suppliers to see the online auction as an opportunity to increase their prices," says Maria. That would have the same effect on the team members' eyes as chopping onions.

Other examples include setting up lots in a way that propels competition, including a smart mix of high-volume items together with those items that suppliers don't like. In this situation, the procurement team needs to ensure that there is high interest in each lot and that suppliers bid on every item in each lot.

RFQs (online auction included) is a powerful tool to obtain savings with the correct preparation. The team can issue an RFQ without a previous RFP or RFI.

The use of the different RFXs depends on how ready the company is to go to market. If the company doesn't know much about the options available out there, everything starts with an RFI to then move to an RFP and to a potential RFQ, if the company wants to use an online auction. In other cases, the company may decide to go with an RFQ in the first place or not use it at all.

Once the RFQ finalizes, based on the selection criteria, procurement awards the business to one or more suppliers that may not be the ones who submitted the lowest bids during the online action.

"What!" says Jon maintaining intense eye contact with Maria. "How are you going to award the business to someone that is not the lowest bidder? That's unethical."

"That's what the lowest bidder would argue," says Maria.

"Oh well, duh," says Jon.

"What happens, Jon, is that sometimes companies include a total cost of ownership clause in the RFX package. This means that price is a factor in the selection criteria, but there are other factors too," explains Maria.

"It's unethical," Jon reiterates.

"It is not unethical if this clause is included in the RFX," says Maria.

Although not unethical, this clause can taste like sour cream and onions for suppliers participating in the selection process. To avoid this situation, companies can run an RFP, assess suppliers' responses, and invite the three or five finalists to an online auction where the determining factor is price. The lowest-cost supplier gets the business.

When the RFX finalizes, contract negotiations kick in. But, as the supplier already accepted the business requirements when participating, 70 to 80 percent is already complete. Whoo-hoo!

Other times, without going through the RFXs, the company may decide to negotiate a contract with a supplier, for example, to formalize a relationship when both parties have been collaborating together for quite some time.

In addition, when contracts are about to expire, procurement wants to negotiate with suppliers. "All things contracts are our focus in the next section, after this Chips in Action," says Maria.

Chips in Action

1. Analyze what your current RFXs are and decide what changes you could implement to turn them into cost savings machines.
2. Build a pipeline of future sourcing event opportunities.
3. Create a calendar with the sourcing events based on ROI and implementation ease. "Here, you need to consider if there are commitments in place like legal contracts and POs."
4. Document best practices.
5. Document lessons learned.

Contract Management

Contract management is not sexy. Who likes contracts apart from legal? Often not listed as a priority for CPOs (Chief Procurement Officers), this area can pour on the savings.

Contract management[87] is the process of contract creation, execution, and analysis to maximize operational and financial performance at an organization, all while reducing financial risks.

Contract lifecycle management consists of nine phases:

Contract Creation
Phase 1: Request—This is the initial step. When companies launch an RFP or RFQ, they take the business requirements as part of the scope of work to be included in the contract. The same applies to the pricing worksheet from the pricing section.
This phase of the request consists of identifying contract opportunities and putting together all the documents, like business requirements, pricing worksheets, and answers to the information questions. If the company doesn't go through an RFX process, the documents can be from conversations with the supplier.
Phase 2: Writing—A great tool for writing a contract is the standard templates from our friends in legal, the boilerplate language that the company uses. This standard language includes insurance requirements, payment terms, contract termination and renewal, protection of intellectual property, and confidentiality, among other aspects. These standard templates contain appendices to complete the specifics of the contract.

Collaboration
Phase 3: Negotiation—This is back and forth when the ball switches hands from company to supplier and vice-versa. If the contract is the result of an RFP, there shouldn't be much negotiation around the business requirements, shortening the time to reach the final contract.

Signature
Phase 4: Approval—Depending on the dollar amount, the approval phase can involve several steps. There can be different approval workflows—in parallel or serial—along with templates that facilitate the review and approval of the contract.
Phase 5: Execution—This is when both parties sign the contract. Electronic signatures simplify the process.

Tracking

Phase 6: On-going supplier performance evaluation—This is a relevant area for cost savings.

"We are so happy with the contract done, but we then don't monitor the beautiful metrics we included," says Maria.

"That's so true," confirms Sarah, "we don't track the SLAs—the service level agreements."

"We don't enforce the savings we negotiated," says Maria.

"Or we don't make them pay the penalty fees and we're the ones who got charged back," says Mike.

"If we have a contract, we need to make sure that the supplier meets the deliverables," concludes Maria, transitioning to revisions and amendments.

Phase 7: Revisions and amendments—Changes can happen. For example, the supplier is doing so well that the company decides to award more business to them. In that case, they can have an amendment to the contract covering the new service. It's important to establish a process to accomplish this.

Phase 8: Reports and audit—Audits are important to ensure that both parties comply with the terms and conditions agreed to in the contract and to detect any issues.

Renewal

Phase 9: Contract renewal—This is another area for potential cost savings. "We want to identify the contracts that need renewal and look for cost savings opportunities with the supplier," says Maria. Contract renewal is not a mere administrative task. It represents an opportunity to improve the contract.

"You guys are asking for it, right?" asks Maria.

"We can't wait," replies Mike with cheerful sarcasm.

"Here it is, Chips in Action," says Maria.

Chips in Action

1. Put together a list of all the contracts.
2. Include status.
3. Prioritize contracts.
4. Set an action plan.
5. Hands-on to execute.

"Let's make it happen, team," says Ernesto.

"We have a couple of effective ways to achieve savings: 1. changes to the packaging and or product, and 2. opportunities in procure-to-pay," says Maria. "Next comes changes to the packaging and or the product."

Changes to the Packaging and/or Product

In collaboration with suppliers, a company can identify changes to the packaging—of course, without having an impact on the functionalities or properties of the product—that can lead to savings, for instance, by changing the design or material that can also have an impact on manufacturing and on the shipping cost. This is another example that supports that the holistic approach is the way to go.

R&D or engineering can also make changes to the product that lowers the cost by switching to different ingredients or formulas. The team also needs to consider the regulatory aspects to ensure compliance. Like with the changes to the packaging, it's fundamental that the team adopts a holistic approach.

"And we are completing the baker's dozen with opportunities in procure-to-pay that relates to contract compliance," says Maria.

Opportunities in Procure-to-Pay

Procure-to-pay is the procurement genie for execution. This is because the procure-to-pay process should enforce the use of the suppliers that procurement selects and the contracts that are negotiated.

Per SAP Ariba definition,[88] the procure-to-pay process is the process of integrating purchasing and accounts payable systems to create greater efficiency.

This process has four main stages:
- selecting goods and services
- enforcing compliance and order
- receiving and reconciliation
- invoicing and payment

"Can you cut to the chase and say where the opportunities are?" asks Jon.

"Thank you, Jon. Of course, there are many," replies Maria.

A procure-to-pay process on fertile soil has the catalogs of the suppliers that procurement selects. In that way, everybody buys from the correct suppliers, from the suppliers that give the most value with awesome negotiated savings. "We are talking about TCO here," says Maria.

This helps to reduce the maverick spend—spend outside of what procurement has defined—and to ensure that all the savings on paper get to the financial statements. There are approval flows that channel the purchase requisition based on the requestor and the dollar amount.

The benefits of a solid procure-to-pay extend to sourcing too. When the POs have line item details and the correct commodities codes, procurement can reap enormous benefits for spend analysis and have fun with the RFXs and contract negotiations.

"Okay, guys, you've made it!" says Maria with great enthusiasm. This is the end of the training, and the team has nailed it.

She adds, "Keep in mind how to show the savings to finance and keep rocking! I'm so proud of you all… and Ernesto, thank you so much for this amazing opportunity, sir. It has been beyond thrilled to help the team."

"Thank you so much, Maria, for your hard work, commitment, and dedication," says Ernesto.

"We're going to miss you," says Amy, while Sarah nods.

"Thank you very much for everything, fellow Spartan," says Tim.

"Good stuff, Maria. Thank you," says Mike.

"I guess I will miss you too, Maria," says Jon.

"Thank you, Jon. Your comments and questions made my explanations better, so thank you," replies Maria.

"Wait! Are you gonna leave us just like that? Without any sort of plants or trees?" asks Ruben.

"I have something for you, team! The last and final Chips in Action," says Maria.

"I'd rather have a plant, even a small cactus would work, but since you insist, what is the final Chips in Action?" asks Ruben.

Maria leaves the team with the final Chips in Action that consists of going over Ernesto's top 10 challenges list for Alex's Snacks. With the frameworks and tools from the training, the supply chain and procurement team needs to make sure that they address each challenge successfully. Maria is absolutely convinced that the team will do a fabulous job.

Every time a project or training ends, Maria feels the same way. It was the same bittersweet feeling with The Long Island New York Chocolates or with BeauTeec Cosmetics, and now she has a sour cream feeling with Alex's Snacks. That's how adventures in supply chain and procurement are, with highs and lows, and plenty of loops.

Hey! Here it's Marcia—not Maria—the author of this book. I hope you've found the frameworks and tools presented helpful. That would mean the world, the planet, and the galaxies to me. With the right mindset, toolset, and work, there is nothing that you can't accomplish. If I can do it, you can do it, with a confidence level of 100 percent. Buckle up and enjoy the ride, supply chainers! Adios amigos, until the next adventure!

Endnotes

1. Marcia Williams, *Hidden Gems that can Make Your Supply Chain Shine*, Forbes, 2022, https://www.forbes.com/sites/forbesbusinesscouncil/2022/11/16/hidden-gems-that-can-make-your-supply-chain-shine/.

2. Oliver Wight EAME, *Integrated Business Planning*, https://oliverwight-eame.com/service/integrated-business-planning.

3. River Logic, *What's the Difference between S&OP and IBP?*, https://www.riverlogic.com/blog/whats-the-difference-between-sop-and-ibp.

4. Elena Dumitrescu, Matt Jochim, Ali Sankur, and Ketan Shah, *A Better Way to Drive your Business*, McKinsey & Company, 2022, https://www.mckinsey.com/capabilities/operations/our-insights/a-better-way-to-drive-your-business.

5. Kees-Jan de Korver, *Integrated Business Planning: Embracing Volatility as a Competitive Advantag*, Accenture, 2021, https://www.accenture.com/nl-en/blogs/insights/integrated-business-planning-embracing-volatility-as-a-competitive-advantage.

6. Tim Berger and Dr. Goetz Wehberg, *Deloitte Integrated-Business-Planning*, Deloitte, 2018, https://www2.deloitte.com/content/dam/Deloitte/de/Documents/operations/Deloitte_Integrated-Business-Planning.pdf.

7. Marcia Williams, *Data and Tech in Supply Chain: How to Make Data Actionable with Guest John Ferraioli*, USM Supply Chain, 2021, https://usmsupplychain.com/data-and-tech-in-supply-chain-how-to-make-data-actionable-with-guest-john-ferraioli/.

8. Mark Levy and Brian Higgins, *Is it Time to Blow up S&OP?*, CSCMP's Supply Chain Quarterly, 2022, https://www.supplychainquarterly.com/articles/6238-is-it-time-to-blow-up-s-and-op.

9. Niels, *The Rise and Fall of S&OP*, Supply Chain Trend, 2015, https://supplychaintrend.com/2015/02/28/the-rise-and-fall-of-sop/.

10. Michigan State University, *How Demand Planning Can Improve the Supply Chain*, 2022, https://www.michiganstateuniversityonline.com/resources/supply-chain/how-demand-planning-improves-supply-chain/.

11. Global Supply Chain Institute, *A Guide to Forecasting Demand in a Stretched Supply Chain*, 2021, https://supplychainmanagement.utk.edu/blog/guide-to-forecasting-demand/.

12 *What is Monte Carlo Simulation*, 2022, Youtube, https://www.youtube.com/watch?v=7TqhmX92P6U.

13 Sunil Chopra and Peter Meindl, *Supply Chain Management: Strategy, Planning, and Operation*, Amazon, 2015, https://www.amazon.com/Supply-Chain-ManagementSunil-Chopra/dp/0136080405.

14 Marcia Williams, *Podcast: Marcia Williams on Supply Chain Optimization and Digital Transformation*, Requis, https://requis.com/podcasts/podcast-marcia-williams-supply-chain-optimization-and-digital-transformation/.

15 James Clear, *Warren Buffett's "2 List" Strategy: How to Maximize Your Focus and Master Your Priorities*, https://jamesclear.com/buffett-focus.

16 Lora Cecere, *Demand Planning. When the Answer to two Simple Questions is not so Simple*, Supply Chain Shaman, 2023, https://www.supplychainshaman.com/demand-planning-when-the-answer-to-two-simple-questions-is-not-so-simple/.

17 IBM, *What is Machine Learning?*, https://www.ibm.com/topics/machine-learning.

18 Berkeley School of Information, *What Is Machine Learning (ML)?*, 2020, https://ischoolonline.berkeley.edu/blog/what-is-machine-learning/.

19 Rohit Dwivedi, *How Does Linear And Logistic Regression Work in Machine Learning?*, Analytics Steps, 2020, https://www.analyticssteps.com/blogs/how-does-linear-and-logistic-regression-work-machine-learning.

20 Kathleen Areola, *How to Forecast Sales Using Product Life Cycle*, eFinancialModels, 2022, https://www.efinancialmodels.com/2022/05/31/how-to-forecast-sales-using-product-life-cycles/.

21 Shaheen Mahmud, *Product Life Cycle and S curve*, ResearchGate, 2020, https://www.researchgate.net/publication/347635761_Product_Life_Cycle_and_S_curve.

22 Dr. Chaman L. Jain, *Product Portfolio Optimization – Journal of Business Forecasting* (Special Issue, Institute of Business Forecasting & Planning, 2016, https://demand-planning.com/2016/02/29/product-portfolio-optimization-journal-of-business-forecasting-special-issue/.

23 Dominic Distel, Eric Hannon, Moritz Krause, and Alexander Krieg, *Finding the Sweet Spot in Product-Portfolio Management*, McKinsey & Company, 2020, https://www.mckinsey.com/capabilities/operations/our-insights/finding-the-sweet-spot-in-product-portfolio-management.

24 Dr. Chaman L. Jain, *Product Portfolio Optimization – Journal of Business Forecasting* (Special Issue, Institute of Business Forecasting & Planning, 2016.

25 Dominic Distel, Eric Hannon, Moritz Krause, and Alexander Krieg, *Finding the Sweet Spot in Product-Portfolio Management*, McKinsey & Company, 2020.

26 River Logic, *Improving Product Portfolio Strategy With What-If Analysis*, https://www.riverlogic.com/blog/improving-product-portfolio-strategy-with-what-if-analysis.

27 Gartner, *Trade Promotion Management and Optimization for the Consumer Goods Industry Reviews and Ratings*, https://www.gartner.com/reviews/market/trade-promotion-management-and-optimization-for-the-consumer-goods-industry.

28 Deloitte, *Trade Promotion Optimization*, 2020, https://www2.deloitte.com/content/dam/Deloitte/de/Documents/consumer-business/Trade_Promotion_Optimization_2020_Deloitte.pdf.

29 Deloitte, *Trade Promotion Optimization*, 2020.

30 River Logic, *S&OP Process: What Does it Mean to Integrate with Finance?*, https://www.riverlogic.com/blog/integrate-finance-with-sales-and-operations-planning

31 River Logic, *Demand Shaping: Strengthening the S&OP Process with What-if Analysis*, https://www.riverlogic.com/blog/demand-shaping-and-what-if-analysis.

32 David Luther, *Cost Accounting Defined: What it is & Why it Matters*, NetSuite, 2021, https://www.netsuite.com/portal/resource/articles/accounting/cost-accounting.shtml.

33 Blazer, E., ACCT364–*Journal Entries for a Standard Cost System*. 2021, Youtube, https://www.youtube.com/watch?v=tVI5JFXeOaI.

34 Robert Kiyosaki, *New Rule of Money #2: Learn How to Use Good Debt vs. Bad Deb*, Rich Dad, 2020, https://www.richdad.com/good-debt-vs-bad-debt.

35 Simon Constable, *How the Enron Scandal Changed American Business Forever*, Time USA, 2021, https://time.com/6125253/enron-scandal-changed-american-business-forever/.

36 Balanarayanan V, *Purchase Order Accruals–Execution with SAP S/4HANA*, SAP, 2022, https://blogs.sap.com/2022/03/20/purchase-order-accruals-execution-with-sap-s-4hana/.

37 David Luther, *Cost Accounting Defined: What it is & Why it Matters*, NetSuite, 2021

38 Abby Jenkins, *ABC Analysis in Inventory Management: Benefits & Best Practices*, NetSuite, 2020, https://www.netsuite.com/portal/resource/articles/inventory-management/abc-inventory-analysis.shtml.

39 Abby Jenkins, *ABC Analysis in Inventory Management: Benefits & Best Practices*, NetSuite, 2020.

40 John Tunstall, *Production Planning Vs. Master Scheduling*, LinkedIn, 2015, https://www.linkedin.com/pulse/production-planning-vs-master-scheduling-john-tunstall.

41 Anne Kaese, *The Modern Manufacturer: MRP vs MPS – What, When, and How*, Stoneridge Software, 2021, https://stoneridgesoftware.com/the-modern-manufacturer-mrp-vs-mps-what-when-and-how/.

42 Austin Caldwell, *Economic Order Quantity (EOQ) Defined*, NetSuite, 2021, https://www.netsuite.com/portal/resource/articles/inventory-management/economic-order-quantity-eoq.shtml.

43 Abby Jenkins, *Safety Stock: What it is & How to Calculate*, Oracle Netsuite, 2022, https://www.netsuite.com/portal/resource/articles/inventory-management/safety-stock.shtml.

44 ASCM, AM 2.2 *Inventory Days of Supply*, ASCM Supply Chain Management, https://scor.ascm.org/performance/assets/AM.2.2.

45 ASCM, AM 2.2 *Inventory Days of Supply*, ASCM Supply Chain Management.

46 ASCM, AM 1.1 *Cash-to-Cash Cycle Time*, ASCM Supply Chain Management, https://scor.ascm.org/performance/assets/AM.1.1.

47 ASCM, AM 1.3 *Return on Working Capita*, ASCM Supply Chain Management, https://scor.ascm.org/performance/assets/AM.1.3.

48 ASCM, CO 1.1 *Total Supply Chain Management Cost*, ASCM Supply Chain Management, https://scor.ascm.org/performance/cost/CO.1.1.

49 ASCM, CO 2.3 *Inventory Carrying Cost*, ASCM Supply Chain Management, https://scor.ascm.org/performance/cost/CO.2.3.

50 ASCM, CO 1.2 *Cost of Goods Sold (COGS)*, ASCM Supply Chain Management, https://scor.ascm.org/performance/cost/CO.1.2.

51 ASCM, CO 2.6 *Direct Material Cost*, ASCM Supply Chain Management, https://scor.ascm.org/performance/cost/CO.2.6.

52 ASCM, CO 2.7 *Direct Labor Cost*, ASCM Supply Chain Management, https://scor.ascm.org/performance/cost/CO.2.7.

53 ASCM, CO 2.8 *Indirect Cost Related to Production*, ASCM Supply Chain Management, https://scor.ascm.org/performance/cost/CO.2.8.

54 ASCM, CO *Level-3 Metrics*, ASCM Supply Chain Management, https://scor.ascm.org/performance/cost/CO.3.21.

55 ASCM, *CO 1.1 Total Supply Chain Management Cost*, ASCM Supply Chain Management.

56 ASCM, RL 1.1 *Perfect Customer Order Fulfillment*, ASCM Supply Chain Management, https://scor.ascm.org/performance/reliability/RL.1.1.

57 ASCM, RL 2.1 *Percentage of Orders Delivered in Full to the Customer*, ASCM Supply Chain Management, https://scor.ascm.org/performance/reliability/RL.2.1.

58 ASCM, RL 2.2 *Delivery Performance to Original Customer Commit Date*, ASCM Supply Chain Management, https://scor.ascm.org/performance/reliability/RL.2.2.

59 ASCM, RL 2.3 *Customer Order Documentation Accuracy*, ASCM Supply Chain Management, https://scor.ascm.org/performance/reliability/RL.2.3.

60 ASCM, RL 2.4 *Customer Order Perfect Condition*, ASCM Supply Chain Management, https://scor.ascm.org/performance/reliability/RL.2.4.

61 ASCM, RL *Level-3 Metrics*, ASCM Supply Chain Management, https://scor.ascm.org/performance/reliability/RL.3.47.

62 ASCM, RL *Level-3 Metrics*, ASCM Supply Chain Management, https://scor.ascm.org/performance/reliability/RL.3.28.

63 ASCM, RS 1.1 *Customer Order Fulfillment Cycle Time*, ASCM Supply Chain Management, https://scor.ascm.org/performance/responsiveness/RS.1.1.

64 ASCM, RS 2.1 *Order Cycle Time*, ASCM Supply Chain Management, https://scor.ascm.org/performance/responsiveness/RS.2.1.

65 ASCM, RS 2.2 *Source Cycle Time*, ASCM Supply Chain Management, https://scor.ascm.org/performance/responsiveness/RS.2.2.

66 ASCM, RS 2.3 *Transform Cycle Time*, ASCM Supply Chain Management, https://scor.ascm.org/performance/responsiveness/RS.2.3.

67. ASCM, RS 2.4 *Fulfill Cycle Time*, ASCM Supply Chain Management, https://scor.ascm.org/performance/responsiveness/RS.2.4.
68. ASCM, RS 2.5 *Return Cycle Time*, ASCM Supply Chain Management, https://scor.ascm.org/performance/responsiveness/RS.2.5.
69. ASCM, RS *Level-3 Metrics*, ASCM Supply Chain Management, https://scor.ascm.org/performance/responsiveness/RS.
70. ASCM, *Introduction to Processes*, ASCM Supply Chain Management
71. ASCM, RS *Level-3 Metrics*, ASCM Supply Chain Management.
72. ASCM, RS *Level-3 Metrics*, ASCM Supply Chain Management.
73. ASCM, RS *Level-3 Metrics*, ASCM Supply Chain Management.
74. ASCM, RS *Level-3 Metrics*, ASCM Supply Chain Management.
75. ASCM, EV 1.1 *Materials Used*, ASCM Supply Chain Management, https://scor.ascm.org/performance/environmental/EV.1.1.
76. ASCM, EV 2.1 *Renewable Materials Used*, ASCM Supply Chain Management, https://scor.ascm.org/performance/environmental/EV.2.1.
77. ASCM, EV 2.2 *Non-Renewable Materials Used*, ASCM Supply Chain Management, https://scor.ascm.org/performance/environmental/EV.2.2.
78. Robert Rich, *The Great Recession*, Federal Reserve History, 2013, https://www.federalreservehistory.org/essays/great-recession-of-200709.
79. Jim Tankersley and Keith Bradsher, *Trump Hits China with Tariffs on $200 Billion in Goods, Escalating Trade War*. The New York Times, 2018, https://www.nytimes.com/2018/09/17/us/politics/trump-china-tariffs-trade.html.
80. *Accurately Convert PDF to Excel*, PDF Tables, https://pdftables.com/.
81. *Every tool you need to work with PDFs in one place*, iLovePDF, https://www.ilovepdf.com/.
82. Dwight D. Eisenhower, *In preparing for battle I have always found that plans are useless, but planning is indispensable*, BrainyQuote, https://www.brainyquote.com/quotes/dwight_d_eisenhower_164720.
83. *Supply Chain Network Optimization: What You Need to Be Successful*. River Logic, https://www.riverlogic.com/blog/supply-chain-network-optimization-what-you-need-to-be-successful.
84. *Multi-Channel Fulfillment Fees*, Amazon Supply Chain, 2023, https://m.media-amazon.com/images/G/01/mcf-new-collection/pricing/Amazon_MCF_Fulfillment__Storage_Fees_1_19_23_finalv2.pdf.
85. *Top Supplier Diversity Programs Broaden Value Proposition To Drive Increased Market Share*, Other Revenue Opportunities, The Hackett Group, 2017, https://www.thehackettgroup.com/top-supplier-diversity-programs-broaden-value-proposition/.
86. Milan Prilepok, Shelley Stewart III, Ken Yearwood, and Ammanuel Zegeye, *Expand Diversity Among your Suppliers—and Add Value to your Organization*, McKinsey & Company, 2022, https://www.mckinsey.com/capabilities/operations/our-insights/expand-diversity-among-your-suppliers-and-add-value-to-your-organization.
87. Bennett Conlin, *The Fundamentals of Contract Management*, Business News Daily, 2023, https://www.businessnewsdaily.com/4813-contract-management.html.
88. *What is procure-to-pay (p2p)?*, SAP Company, https://www.sap.com/insights/what-is-procure-to-pay.html.

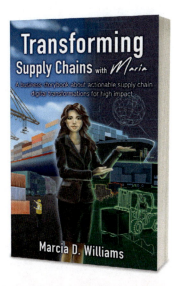

READ MORE!

Transforming Supply Chains with Maria is about how supply chain digital transformations can help you deliver results with a high ROI.

Maria, the main character, will take you through the good, the bad, and the ugly of the following supply chain digital transformations:

- Strategic Sourcing
- Enterprise Resource Planning (ERP)
- Integrated Business Planning (IBP) or Sales and Operations Planning (S&OP)
- Procure to Pay (P2P) process

Read this book, take action, and deliver high impact results!

READER REVIEW

Excellent Book - A must have for anyone who is connected in some way to Supply Chain.

This book is well thought out and well written. I have been in supply chain for 30 years and I am very impressed."

Chris C.

AUTHOR BIO

With almost two decades of experience in finance supply chain adventures, Marcia has had the privilege to work with high-profile companies—many Fortune 500—including Hershey, Lindt Chocolates, Coty, Cummins, and Alcoa. Starting as an accountant in Uruguay, Marcia entered the supply chain and procurement world through Michigan State University, where she also experienced snow for the first time.

Marcia's New York-based business, USM Supply Chain, helps fast-growing consumer packaged goods (CPG) companies achieve greater profits through revenue increase and cost reduction with proven financial and operational frameworks, analytics, and automation. Marcia is a consultant, educator, writer, and speaker. When their three children are not fighting, Marcia and her husband, Jason, are the proud parents of Alex, Emma, and Rebecca (Becky).

Website
www.usmsupplychain.com

Phone
646-814-0696

LinkedIn
linkedin.com/in/marciadwilliams/

Forbes
forbes.com/sites/forbesnycouncil/people/marciawilliams1

TEDx Talk
youtu.be/u_gBJI3LOek

Interested in **Buying 10 or More Copies?**

Reach out to us for our discount schedule and have Marcia speak at your organization.

E-mail:
info@usmsupplychain.com

Phone:
646-814-0696

or submit the form here:
usmsupplychain.com/get-more-copies-of-book

REVIEW REQUEST

Thank you so much for reading my book!

I hope you've found the frameworks, processes, and tools presented helpful. That would mean the world, the planet, and the galaxies to me. With the right mindset, toolset, and work, there is nothing that you can't accomplish. 100%.

I really appreciate all of your feedback and I love hearing what you have to say.

Please take two minutes to leave a review on Amazon so I can make the next version of this book and my future books better.

Thank YOU so much!

Marcia

Made in United States
Orlando, FL
10 April 2024